HOW TO GROW WINGS

The True Story of a
Boy Engineer and the Cold War

by Lyn C. Stafford

EASTERN LAKE BOOKS
Birmingham, Alabama
2025

Other books by Lyn C. Stafford

The Road to Margaret

Expat: Chronicals of an expatriate in Saudi Arabia

For William R. Brosco

Table of Contents

Author's Note vi

Chapter 1. Fortitude and Grace. 1

Chapter 2. Mobile, Alabama 15

Chapter 3. The Blues 27

Chapter 4. Peace That Is No Peace 45

Chapter 5. Home Away From Home 61

Chapter 6. On to Greenland 73

Chapter 7. Days on Ice 91

Chapter 8. Supply Ship 103

Chapter 9. Last Years at Thule 113

Chapter 10. The Cold, Cold War. 125

Chapter 11. A Call from Chrysler. 135

Chapter 12. Serious Stuff 149

Chapter 13. Redstone Comes Home. 159

Chapter 14. Jupiter AM-18 and Miss Baker. 173

Chapter 15. Super-Jupiter 183

Chapter 16. The IGY and Jupiter C 191

Chapter 17. Michoud and the Saturn Program. 203

Chapter 18. Missile and the Grey Goose 213

Chapter 19. The Cuban Missile Crisis. 223

Chapter 20. Trouble in St. Tammany Parish 237

Chapter 21. Missing Boys 247

Chapter 22. Ceremony 257

Chapter 23. Data Center. 263

Epilogue 271

Acknowledgments. 277

Photographic Credits 279

Bibliography 280

About The Author. 286

Author's Note

William R. Brosco is the hero in our story, and like most heroes, he has stories that need to be told. But this is not just his personal story. It is the story of a life lived during the years of and within the context of the Cold War, one of the world's most fascinating eras.

Bill Brosco began tinkering with all things technical when he was a teenager in Sioux Ste. Marie, Michigan, near the end of World War II. In pursuit of his dreams, he seemed to fall into top secret assignments over and over again, through no design of his own; but wherever he was at any given time, he, not unlike many young men in that era, played a vital part on America's side of the Cold War. As a young, conscientious engineer, he was the perfect person to send out into a clandestine, confusing, and usually dangerous world.

Bill's father, who had unfortunately been named Adolph, came from Poland long before World War II. He could only watch from afar as Adolph Hitler destroyed his homeland first, then destroy country after country as World War II raged. It was no wonder he swore to fight Hitler or anyone like him, and he expected all his sons to do the same. As the war continued, Bill's father suffered even in the United States because he, too, was named Adolph.

The younger Brosco knew he had to join the armed services, and almost before he began his adult life, he became his own breadwinner. With hard work and a strong desire to survive, he never had to search for a job. He just fell into them. As he ventured from one project to another, he was caught up in secrecy and controversy among industrialists, countries, politicians and people in high places during the Cold War. Bill's story is, in many ways, the story of that long, Cold War.

To place Bill's stories with chronological accuracy, the author carefully researched history, especially the histories of events during Cold War that so affected Bill's and all our lives, as well as the histories of the Redstone rocket and its descendants. Some of the locations and timing of his stories therefore may vary from the telling, but the lived experiences of Bill Brosco are historically accurate.

The stories in this book are Bill's own stories, stories from his past told in his own words. With the passing of time and because humans do change their perspectives during their lives, the telling of these stories may have limitations. The telling of these stories might vary to conform to history. The author has attempted not only to recreate time and place, but to verify each and every event with research from as many resources as possible. After so many years there are many accounts of the same events.

One thing has not changed: the brutal struggles for supremacy between the East and the West has never ended, and wars closer to home have never stopped. But from the author's perspective, Bill and the United States may have done something right, because there was a time during those years when peace almost succeeded.

The views and opinions in this book do not reflect either official sanctioning or endorsement by any element of the Departments of the Army, Navy, Air Force, or any Defense or civilian groups, including the Chrysler Corporation, involved in those projects during the Cold War.

CHAPTER 1

Fortitude and Grace

On that cold day in December, the sun was about to set on Sault Ste. Marie as the youngest employee of Northern Electric Company was standing outside, on a ladder, trying to fit a wire through a brick wall into the top floor of a house where an old man lived with his wife. He managed to stuff the wire through the hole he'd just drilled and was about to pack up his gear, but he wanted one last look from such a height. Off to the west, beyond his house on Portage Street, lay the St. Mary River, and beyond that stood the old lighthouse where he'd been born.

He saw it. He'd biked there a lot that summer of '46, and memories of that lighthouse helped block out his life at home, for a little while. There wasn't much left of that old lighthouse now, but then again, there wasn't much left for him at home, either.

Those months felt like years now. He just wanted a little time after school, after he graduated, to persuade Mr. Marriott that he could have that old lighthouse working again in no time and stop the State of Michigan from tearing it down. He could and was probably the only person who would bring that old lighthouse back to life.

But there would be no after school. His father, Adolph, had made that clear.

Why was it that Mr. Marriott always gave him jobs out in the cold? And why old Colonel Grable? The Colonel was the Company's best client, rich but kind of strange, notorious for wanting the latest and best of everything, a stickler for punctuality. Today, in the dead of winter, he wanted Northern Electric's newest and most famous AC/DC table radio, called the "Baby Chimp," installed on the second floor. It was his Christmas present for Mrs. Grable.

Not until Bill took all his gear up the stairs to the landing and began unloading boxes did Bill discover something Mr. Grable had failed to tell anyone back at Northern Electric: there wasn't any wiring upstairs. That was now Bill's problem.

The Baby Chimp was more reliable than its predecessors. It was an innovation that promised Americans they would no longer have to wait forever to get their news. The Brosco family had heard about the bomb hitting Hiroshima a day late: the main lines of communication had failed. Radio stations went off the air in Japan, and the telegraph even stopped working. People hundreds of miles away from the site where the Enola Gay dropped that bomb saw the smoke cloud rise like a mushroom in the air long before they even began trying to find out what had happened. Sixteen hours later, after communications had been restored enough to send President Truman's crackling voice through the old radios, most of the world heard the story above static, and journalists raced each other to get the whole story right the first time. Japan's Morse code operators were the only truly reliable messengers. Now Mr. Grable would get breaking stories loud and clear through a fancy new rainbow grill—that is, if Bill had connected the wire right.

Glimpses of the river distracted him now and again, and he would lose his concentration, mostly only a fraction of a second when the sun's rays sparkled on a wave in the distance. He'd long ago learned he could bring his concentration back if he bit his lips, and he realized he'd been biting them ever since he first got up on that ladder.

Now he'd finished, so he took stock of what he'd accomplished. He had been in too much of a hurry to get his ladder up at the right angle from the very beginning of the project, because he found himself hugging the wall. That ladder was way too tall to have positioned it this close to a wall. He knew better. He'd hugged that wall every time he looked out over that river and every minute he spent calculating how far it was to the old lighthouse. The ladder was now the issue at hand. He had to concentrate on steadying himself as he put the wire and his tools away without dropping them, especially since some of his classmates were standing around down there watching him. He knew he was great entertainment for the after-school crowd.

Temperatures in Sault Saint Marie, the Soo, at this time of year could go as low as 30 below, so time was of the essence. He gripped the cold rungs so hard he could hardly get the wiring back in his kit, but at least he could still feel his hands, gloves or no. Just five more minutes. He would finish this thing in five minutes exactly. Clouds rolled ominously in from the west and would bring snow for sure. More snow meant more ice. Both meant trouble.

Old man Grable had hardly been able to contain himself he wanted that radio so badly. Even so, Bill doubted Mrs. Grable would ever even have a chance to turn those dials, but the Grables would have a good Christmas.

For Bill, Christmas and those dark days of winter signaled only one thing: the beginning of the end of what began as high hopes at the start of his senior year in high school and was about to end before he could finish. A future, which he'd hoped might be working for someone like Mr. Marriott, seemed more and more unlikely. Right now, he had to return the ladder to Mr. Marriott then start the long walk home. He would be late. His mother would be angry. He had no choice: it might take him another hour before he could even begin his chores at home.

By then the water in the lake by the farm would be iced over. He could only pray that the ice wouldn't be too solid to keep him from knocking a hole in it. He didn't need much, just enough of an opening to draw water for the animals. Anything left from morning would already be frozen over. He tensed even more, praying the ice on the lake would be thin and the chickens and cows would already be sheltering in the barn.

Mr. Marriott was a kind boss. Bill was the only employee he had that was still in high school. Bill was the lucky one. His boss didn't trust anyone but him, especially when it came to old Colonel Grable. Bill had hauled that thing all the way up those steep steps to the second floor in the house before he realized there was no electricity up there, much less an antenna. He'd had to wire the whole damn thing, which meant going back to the shop for a longer ladder before he could even get to the second floor of that old building, but he'd done it in record time, and now he was finished.

He felt a moment of pride standing up there on top of that ladder, but he had to come down. The other kids were down there kidding around with nothing to do, and he was looking at his handiwork and pondering how it all worked, how a common wire could just grab energy from out of the air and unscramble signals from ever further distances. Nobody else had anything like it, and very few in the world even knew it existed much less how it worked. It worked because it had a new radio tube. He knew that. Mrs. Grable could boast of having the latest thing.

"Why don't you come inside? It's cold out there," came Mrs. Grable's voice from inside.

"Because the work is out here," said Bill under his breath, then "Thank you, ma'am, but I've gotta finish up before it's too cold to work outside."

He really didn't want to go into that house if he could help it; it would be hot and musty, plus the old lady just wanted someone to talk to, and it would cost him time he didn't have. It didn't help that some of his fellow seniors stood around on the ground below his ladder, looking up, watching every move he made, curious. All of those kids walked home together every afternoon when school let out, and Virginia was among them. She was the one who always hung back, sort of waiting for him, he thought.

"Whatcha doing now?" One of the fellas shouted up at him, just to annoy.

"I'm about to finish," he yelled down.

"Yeah, sure. Ya'know burglars leave ladders leaning on windows in upstairs rooms."

He could hear girls giggle; he knew they were ribbing him, but it wasn't funny to Bill. He was damn lucky to have that job and didn't want to screw it up. He clamped his tongue even tighter between his teeth so he wouldn't act like a fool and fall off the ladder. Still, he couldn't help but look down to be sure Virginia was still there. She was.

"I'm coming down. Don't let me knock any nails or tools down on you."

The little crowd backed off but didn't leave. Watching Bill climbing up and down a ladder was the most entertaining thing they could do that day.

As he adjusted his tool belt, he paused long enough to look through the trees again in the direction of the old lighthouse, perhaps for the last time.

He could almost hear his mother talking. He knew what she would tell him—"pride goeth before the fall." She'd done everything she could to turn him toward the cloth, to turn him into a priest, and she was definitely down on pride. But any dream his mother might have had about his becoming a priest was history, no matter whether Bill liked the idea or not. His father was sending him away.

All the way home he rehearsed what he might say to talk Adolph out of it. All he'd ever wanted was to finish his senior year and work for Mr. Marriott until he had enough money to leave home forever and see the world or, and he could only dream, enough to go to college.

When his family lived in that Coast Guard lighthouse near the Sioux locks, he was still a child, too young to accompany his father out on the docks, or at least his father—whose name was Adolph—never let him. But there was one day when not even his father wanted him to stay at home. That was the day his mother was sick in bed. Only now, looking back at that day, did he realize that his father, the lighthouse keeper, had no choice but to take him along to work on the docks. He was the only one still at home; both Jack and Adele were old enough to go to school.

That had been the finest day in his childhood memory. He'd been rejected so many, many times before. He had even tried holding on to his older brother's hand, but his father had brushed him off. Jack would never have let him tag along anyway, if he hadn't been in school. Jack, too, brushed him off. If Adolph frowned,

Jack frowned. The two were tight as ticks. Jack was the family football hero.

Those few days with his father those long years ago between Jack's first day of school and the day his mother recovered were his heaven on earth; he spent those joyful days at the lighthouse watching Adolph carry out the never-ending tasks of a lighthouse keeper, and he took it all to heart. He asked Adolph to tell the story again about his childhood in Poland—the story about how Adolph and his cousin ran away from Poland. All the family knew the story.

Before 1920, Soviet Union troops invaded Poland and captured Polish boys along the way to serve as Russian soldiers. The USSR's goal was to take over independent Poland and use Polish children to do their fighting for them. The Soviet Union had no intention of stopping with Poland, though. They wanted everything, all the way to the borders of Europe and probably beyond. The Russians wanted Polish boys who could fight, so they began grabbing boys as they marched through Poland. Any Polish boy who resisted conscription signed his own death warrant; he would be shot then and there.

The two boys were barely teenagers when they ran anyway, but they were the lucky ones. They made it to Canada, and there they joined the Northwest Mounted Police. That was when they lost their Polish name. If the immigration officials had been more patient, they would never have had the name Brosco. That happened because an immigration official, whose job it was to enter Adolph's name into the books, couldn't understand them. He was impatient..

"Your name. Give me your name. Speak English! I can't understand you." Adolph tried again and again. He even spelled the name he was given at birth, but spelling using the Polish alphabet was useless. The official didn't know their alphabet. Finally, Adolph looked out the window at a solid brick wall. It was some kind of warehouse with the words "HILL BROS. CO." painted in large white letters on the dark brick. Adolph pointed out the window, and the official wrote B-r-o-s-c-o on the form. That was that. Adolph and all of his family would forever after be known as "Broscos."

As each of the boys married, he left the Mounted Police and took his family to the States. and the first thing Adolph did was to join the United States Coast Guard. The Coast Guard made him a lighthouse keeper. That was how come a very small Bill was holding tight to his father's hand that day as the two watched magnificent ships make their way across Lake Superior near the St. Mary's River and on toward the Soo Locks and Lake Huron. It had been a glorious day, one he would remember forever because on that day, when he asked questions, his father answered them. Not only that, Adolph let Bill go inside and touch the controls and look at the log books. Bill held onto every word his father told him that day.

When the first ship's horn blasted, Adolph took Bill inside the lighthouse and bent down close to Bill's ear and pointed out the radio where they could hear the captain of the ship talking to them. Then, from off the rocky coastline, they could see a boat approaching under a sky as blue and clouds as white as the ones in the stained glass window of the church. The wind had blown steadily but gently from the lake that day, and all was well.

When the ship grew closer, Bill could see a man standing on the bow and smoke billowing out of the smokestack behind him. Then the ship's horn sounded three long blasts, and Adolph raised his megaphone to his lips.

"Name?" he called.

"SS Superior City, Cleveland Shipbuilding," the man called through his megaphone.

"Destination?"

"Whitefish Bay."

"Origin?"

"Ecorse, Michigan." Adolph wrote it all down in his log book and radioed it ahead.

Bill wanted to know everything. "Why is the captain outside? Why isn't he driving the boat?"

Adolph took time to explain. "His engineer is driving the boat, son. He's got to be on deck to tell his engineer how to steer his ship through the locks."

Bill understood that his father needed that information to decide whether to allow the ship to proceed or not. Bill would ask about that later.

The captain shouted his answers even after his ship began moving ponderously away from Adolph toward the locks. Bill tugged at his father's sleeve, but Adolph finally stopped answering.

"Request to enter the Soo" the voice boomed, as rough water lapped against the rocks. It was time. The captain slowed his engine just enough to approach the locks and go on his way south. Adolph told Bill to hush and listen so he could finish writing in his log book. Bill asked one more question and never forgot its answer:

"What's a captain, Dad?"

"Well, he's the Chief Potato on the ship. He's the one who knows how the ship works. He's the one who has to know everything," Adolph said as he closed his log book. It was clear to Bill then and always: he, too, wanted to know everything. And his desire to know everything began with having to figure out how to install this new-fangled radio on the top floor of an old brick house on a frigid winter day.

Before he could start down his ladder, he had to be certain everything was working as it should. Sweat formed on his forehead, but he couldn't let go of the ladder to wipe it away. The sides of the ladder jerked just enough to remind him that his ladder wasn't leaning quite right. It was too vertical for its length. "Pythagorean theorem," he thought as he leaned into the wall and hugged its rungs. His father would have cussed him out if he'd seen how he'd positioned that ladder. It was not the steadiest of all perches today, but it was the only one that would work.

"Turn it on, Mrs. Grable!" he hollered through the open window. He heard her walk across the room.

"It's on, son."

He screwed the outer panel on. It was live. That was it! His work was done.

He might still be home before dark. He made it down to the bottom rung of the ladder and gently put his right foot down on the brick walk, testing to be sure the ladder wasn't about to fall, but in the process stepping on a corner brick. It gave under his weight, and his skinny leg buckled. He stumbled on the frozen ground doing his best gymnast's roll and stood up, checking himself briefly.

"See you tomorrow," he told Virginia, who was the last of the hangers-on. He really liked her, but he had no time to talk today. Mr. Marriott would be waiting to close his shop.

"No you won't! No school tomorrow. It's Saturday, and anyway we've got a Christmas party to go to. You coming?"

He had forgotten. He had actually forgotten. He'd thought of nothing for weeks other than that he would turn eighteen on Christmas Eve and was supposed to graduate with his class in June, but his father had thrown up a roadblock. Adolph worked for the government now as a recruiter for the armed services, and he had already signed Bill up for the draft. Bill still couldn't believe his father actually did it.

If he were to bring in enough money, give it all to his mother, would his father change his mind and tear up those enlistment papers? Otherwise Christmas Eve, which was his eighteenth birthday, would mark the end his life. He really would have to join the Army.

He gathered his toolbox and collapsed the ladder, then called out to let the old man know his job was done. As soon as he collected the money, he collapsed his ladder and shouldered it and was about to begin his long, cold walk to Mr. Marriott's shop, when he realized Virginia was still waiting for him.

"I'll see you at the Christmas party!"

The last of the blue sky deepened into one wide, black, ominous cloud against the eastern horizon; behind him, to the west, there was still an orange glow. As he watched the cloud mass forming overhead, a squadron of Army Air Force planes flew in V-formation, wing to wing, toward their base. Bill saw them just fine, but when he covered his good ear, he couldn't hear them at all … still deaf on the side where his mother hit him.

It had been a week since she'd come up behind him and slapped him on the ear, and he should be hearing by now. But then again, she'd hit him so hard he'd almost blacked out. When he asked her what he'd done, she never answered. Still, he was young; he thought his hearing would come back.

On that same night, Adolph came upstairs into the room. Bill was testing his hearing.

"Your mother and I have decided it's time for you to leave home. You already know I've registered you, and you are going to join the Army when you turn eighteen."

This was no longer Adolph's war. Why was he sending his son away? Why did his mother agree? He would puzzle over their reasons for the rest of his life.

On December 24, 1946, he would turn eighteen, and he would be gone by January 1, 1947. But as any seventeen-year-old, he had other dreams for his future. Hitler was definitely dead. He died in 1945, so why wasn't the war over? They listened to the radio every night, and they heard when the bomb dropped in Hiroshima and when Hirohito surrendered; but they heard nothing about the war ending. Adolph always called the family in when there was an announcement coming over the radio. So why were all those bombers still flying around? Then again, perhaps the war really would be over in January.

For Adolph, Hitler's death meant something else. That should have stopped people from mocking his name, and without all the aggravation, he might go back to the Coast Guard. After all, Jack was now in the Coast Guard, and if both of his two oldest sons were in the U.S. Armed Forces, perhaps his neighbors might accept him not just as a loyal American citizen but as the father of sons who also served.

General Eisenhower told the nation that even though we'd been victorious, the United States had to keep its guard up. As long as missiles were in the hands of nations like the Union of Soviet Socialist Republics, which was of course Russia, he would keep his Army Air Corps planes flying. Ike called it his "Army of Occupation." The world would just have to wait for him to declare the war officially over. Those pilots Bill could see flying overhead were still at war.

Bill steadied himself as he walked toward the shop, but could not take his eyes off the bombers. He yelled up at the planes and tried to shake his fist at the sky without dropping the ladder—as though those pilots in their cockpits could hear him anyway.

"I'm gonna fly, fellas!" He yelled at the departing planes. "I'm gonna fly!" then he added almost as an afterthought, "I flew a Piper once!"

It was true. He wasn't kidding. The father of his best school buddy, Jim, worked for the Piper Corporation and bought himself a Piper, which he sometimes flew to work. The Piper factory was a long way away.

There were times Jim's father took his boys and Bill out to the airfield to teach them all the instruments and let Bill take a turn in the cockpit. Then one day Jim's dad let his boys and Bill take the yoke and fly! Jim's father was with them all the way.

As Bill watched the wings disappear over the Soo, he cocked his head, trying hard to hear the bombers' familiar drone as they disappeared into the distance. He could hear with his left ear, but with his right, he could barely hear himself shout much less hear retreating planes.

In the weeks since his mother hit him, he'd tried to figure out why. He didn't do anything to deserve it, but guessed one of his five younger siblings probably did. One of them probably blamed him for something, most likely Mort. Mort hadn't even been born when they lived in Mobile. He was one of Bill's five younger siblings born after they moved back to the Soo. Walter was okay; Eve was just a nuisance because she followed him around everywhere he went; but Mort was just plain bad.

The light from the shop still shone up ahead. He was in luck; he was going to have to tell Mr. Ray Marriott sooner or later anyway. The bell over the door tinkled when he entered the office, and his boss looked up from his desk in the corner.

"All done?"

"Yessir," said Bill. He hesitated a moment trying to decide how much or even whether to tell his mentor about not finishing high school. He knew from the beginning not to ask his boss for advice. His boss hated people who bothered him with too many questions; he thought them incompetent. He was big on people figuring out things for themselves.

Bill also didn't want to appear ungrateful. He'd learned so much from this man after school—installations and repairs in houses and shops, testing for and solving refrigeration problems in air conditioners and issues with appliances, wiring and pipelines. He had to make his announcement now or be quiet.

"I think the Grables are going to have a great Christmas, Sir, and I got 'em all fixed up. You've taught me so much these last months, and I'm very grateful, but..." He thought not more than a second or two before he finished. "I think I may have to quit, Sir, because I may not be here next year. I'm not absolutely sure about that, though; it depends on my father. He wants me to join the Army."

Bill hung his head and told Mr. Marriott the whole story, about how his father had told him he had to join the service sooner and not later, that he was supposed to leave right after Christmas— after his birthday. But, he added quickly, he was going to talk to his father again and hoped he would change his mind and let him graduate first. His after-school job was a great privilege, he told his boss, and he knew some of the other boys envied him. He'd been voted the vice-president of the senior class!

"Sir, of all the things I've done in high school, like working out at the gym and singing in the chorus and working on the school council are not near as interesting as what I've learned from you. I love working with electricity and radios and air conditioning; this job may be the best thing I've ever done."

It was true. No other boy, indeed no other man he knew had this much training on these latest American marvels.

"I want you to know how much I appreciate everything you've done for me."

"What? Boy, you're the best employee I've ever had. Why would you quit?"

"I don't have that much time left. I'll be eighteen on Christmas Eve, but I'm sure going to try. You know, my dad's the Navy Recruiting Agent for the whole county. I can't predict what he'll do."

Mr. Marriott said he knew Adolph and how stubborn the man was. "You need to finish high school, and I need a helper. Is that what your father wants?"

"Yes Sir! I mean, no Sir. He and my mother have already signed me up."

That was too much for Mr. Marriott. The man was going to send this promising boy away? Not let him finish high school? He crossed himself, then waved Bill out with a very Catholic 'bless you' gesture.

""I guess all I can do, Bill, is wish you well," and Mr. Marriott sat quietly for a minute. "Keep in touch, Boy."

The next day at school, Virginia asked him if he would let her write his legacy for the 1947 yearbook, and he told her yes that would be fine. She couldn't tell him what that legacy would be, because she hadn't yet decided.

That was when he had to tell her he might not graduate, but how could he explain the "why" to Virginia? He'd told his boss everything, but for some reason he couldn't explain it to her. All he could say was how grateful he'd been to Mr. Marriott for giving him an after-school job.

CHAPTER 2

Mobile, Alabama

When the little family lived in Mobile, Adolph traveled most of the time, and Mama Evelyn stayed behind with her three children. Adele was twelve years old, Jack was ten, and Bill was almost nine.

Evelyn enrolled them all in the Catholic school and told them all to pray for their father every night so he would come home. Bill prayed faithfully. He knew that if Adolph were to come home more often, his father would understand his mother's loneliness and how hard it was to be a Yankee in the South. Maybe then he would take his family back home to Sault Ste. Marie.

Bill hurt for his father every time someone slammed his father's name, Adolph, the name of the most hated man in the world. No matter how much Bill bragged to the other kids that his father was in the Coast Guard and away at gunnery school, it made no difference to kids in the South.

The insults began long before they moved to Alabama. When Adolph brought his new, young Canadian wife to the States, the same year he became an American citizen, his name—along with all the others who had just been naturalized—was published in the newspaper. Unfortunately, that newspaper came out just when Adolph Hitler began murdering people in Poland. Even though Adolph's last name was Brosco and not Hitler; even though his blonde hair came from Polish ancestors and not from some supposed master race, to some American ears his

Germanic accent was indistinguishable from that of the Fuehrer. His neighbors now either ignored Adolph or mocked him behind his back, and his family suffered the same taunts and assaults that some Americans directed toward anyone who might have a connection to Germany.

That was why Adolph flaunted his patriotism, that was why he became a gunner's mate for the U.S. Navy's Coast Guard, that was why he took a government job in Michigan recruiting sailors for the Navy, and that was why he accepted an assignment at the Port of Mobile, Alabama.

Bill remembered days his father came to town and kids would gather around his new car and throw rotten tomatoes and eggs at it. They called him names like "Dirty Kraut," and Bill remembered how his father would get out of his car and slam the door to make them scatter. The street kids were actually afraid of him, but boasted about having the courage to confront him.

Bill turned ten while they lived in Mobile. He and his mother and his two siblings lived in a small apartment about a mile from school, and even though kids sometimes poked fun at them, too, his mother told them to ignore the insults. Nevertheless, Bill always had a feeling that the day would come when something bad would happen, and it did.

Walking home from school one afternoon ahead of Jack and Adele, Bill saw two students up ahead, waiting for him! Being the new kid, he would be happy to have anyone walk home with him, but he knew his mistake as soon as he rounded the corner. Those kids were waiting to beat him up.

"Go home you damn Yankee or come over here and fight." Bill walked past, but they circled around behind him, shouting "Yankee go home! "We don't want Krauts or Yankees here!"

Those were the last thing he heard before the two jumped him. He tried to fend off the blows, but the fat kid pushed him down on the sidewalk, and the skinny kid beat him up good and proper. They'd gotten him pretty bloody by the time Jack and Adele rounded the corner. Bill had never been so glad to see Jack in his life, and Jack didn't let him down.

Jack sprang to Bill's defense and pushed the fat kid off, grabbed the other one by his hair, and punched him—hard. Bill struggled to stand up; his nose was bleeding and blood was all over his shirt. Jack had blood on his shirt, too. The fat kid just stood back and watched as the skinny kid crumpled under Jack's blow.

"I ain't done nothing," the fat kid kept saying. "It weren't me."

When the skinny one got his bearings, he strutted back down the road as though he'd been given a few medals for taking a licking, and the fat one followed.

Jack helped Bill to his feet. "Hold your nose. Like that," he said as he pinched Bill's nostrils shut. "You gotta learn how to punch out guys like that. I'm gonna give you a fighting lesson when we get home. There's one thing you gotta remember. The most important rule of all is you never, ever, let the bad guy get behind you."

Bill would remember that rule for the rest of his life.

When they got to the apartment, Jack and Adele tried to sneak Bill in, but he dripped blood all over the floor, and the path of blood drippings did not escape their mother's eagle eye. She followed Bill into the bathroom and turned him around to face her. She was mad at him and shaking her finger at him the whole time she talked. He wanted to tell her he would clean it all up, but she kept shaking, and he couldn't get his words out.

"Boy, you turn around right now and go back outside and find that boy who hit you, and you hit him on the nose as hard as you can. Hard! There!" she said as she touched the part of his nose that lay just above where blood was gushing out. "I mean it."

Bill couldn't believe what he was hearing. She was the one who always preached about not starting a fight. She was the one who wanted him to be a priest. But she wasn't the only one who wanted him to be a priest. Even Mama Evelyn's sister, who was a nun, couldn't wait to see him in a priest's cassock. Bill got a double dose of that message from family and Mobile's Catholic schools.

As for Skinny Boy, Bill had to do what he was told, wondering all the while what had become of the priest idea. He went out alone this time, without Jack, to hunt for his assailants, his hands trembling. Those boys towered over him and he could still feel the blows. But it hadn't been a fair fight. They held him down and he couldn't do anything to defend himself. Then in the next moment, he saw them up ahead, laughing and talking. First, he felt anger, then the anger turned to rage and gave him courage.

He obeyed his brother's instructions and followed the Skinny One right in step. When those kids looked back, they saw a determined kid coming at them and, to Bill's surprise, they set off running! He watched them until they disappeared around the corner before he turned to go home. Only then did he see Jack, not far behind. He must have followed him.

"I just looked tough," Bill told his brother. It never occurred to him that the Skinny Guy might have beat it because Jack showed up. The two boys walked home together.

"What happened?" his mother asked later.

"They ran away," Bill told her, still amazed. He had won the battle without hitting a lick.

The next time Adolph came home, he brought a huge, professional punching bag so his boys could practice their punches, and he made them take boxing lessons. He even made Jack and Bill fight each other and after that, Bill knew for sure he would never make it as a priest.

As summer came to a close, Adolph announced they would move back to Sault Ste. Marie in time for Bill to start high school. For the rest of the summer, because Adolph thought they were old enough to know, Adolph told them all the stories about what happened to the rest of their family in Poland and why it was important for them to know.

He had left Poland after World War I, in 1920, but since then, Poland and Germany and the Soviet Union had signed a non-aggression pact promising they would not invade one another. He'd had no reason to think there would be more trouble. Then word came that the Soviets had broken that pact, but Germany had not. Polish citizens didn't worry too much about German aggression against Poland because there were Germans who were their neighbors!

Then, after Adolph emigrated to Canada from Poland with his older cousin, the one who talked him into joining the famous Northwest Mounted Police, and after Adolph donned that uniform with pride, he won the hand of a young girl named Evelyn Porter. After the two married, he left the Northwest Mounted, and the couple moved happily across the border to Chippewa, Michigan. As soon as Adolph and Evelyn became citizens of the United States, Adolph joined the Coast Guard and forgot about Poland.

Then, in 1939, Germany attacked Poland. Bill was ten years old. The United States failed to respond, failed to help Poland, and Adolph became an angrier man than he had been before, even though he understood it had been the kind of surprise attack that no country could have been prepared for. Poland was a small country, and there was a very big ocean between Poland and the United States. Still, when it happened, all the adults who lived nearby hung out just so they could listen to its radio. They'd lived through that last war and didn't want another one.

In the beginning, Germany accused the Polish people of attacking those Germans who lived in Poland, and Germany claimed it had to defend its fellow Germans. But that was not what happened. Polish families in America began receiving long-delayed letters—some smuggled into the United States from families back in Poland—telling an entirely different story. The Polish people wanted the world to know the truth, and little by little, letter by letter, they learned what did happen.

Hordes of Germans had crossed the border into Poland, pretending they were Polish. The locals knew they were neither Polish nor Polish Germans, so just watched what they were doing with curiosity. After all, they had a treaty with Germany. The invaders spoke Polish and wore Polish uniforms, but they were not Polish. They carried German guns and set up camp near the border. For days the locals watched, but on the day those fake Polish troops left their camps and invaded Adolph's old neighborhood and began shooting, they were horrified.

The Germans continued to broadcast to the world that Polish soldiers had done the killing and Germany was only protecting itself. Adolph's neighbors back home told Polish people in the United States that it was the other way around. And soon after, the Soviets attacked, too.

The truth was that Soviet Union and Germany signed another non-aggression agreement between themselves, which they called the Molotov-Ribbentrop Pact, on the same day those Germans attacked. In that pact, Germany told the Soviet Union it could have part of Poland, "their area of influence," which was the north-eastern part of Poland, and the Germans would take the rest.

Hitler ignored that first non-aggression pact completely, and turned his National Socialist Party into a Nazi vicious army that invaded Poland and began killing Polish people. Nobody in Germany, except a few Germans in the other party, objected. Hitler ordered his troops to "kill without pity or mercy all men, women, and children of Polish descent or language." There was never time to mount a defense.

Only two weeks later, the second attack came. The USSR attacked Poland from the east. By December 1941, more than four million Soviets were dead and Germany had captured three million prisoners.

Whenever Adolph read those letters or told that story, he grew teary-eyed. Europe, at first, blamed Poland, then they blamed Hitler, and many Americans blamed people like Adolph because he bore that name. It was no surprise to anyone in the Soo that Adolph volunteered to be the Naval recruiter for the entire district, and really no surprise that Adolph wanted Bill to join the military as soon as he turned eighteen.

How could anyone confuse his father with a Nazi? Some didn't. Mr. Marriott didn't. He knew the whole story about Germany invading Poland, and because he knew Adolph; he wasn't too surprised when Bill said he had to leave school to serve his country.

A few days before Christmas, Bill went back to see Mr. Marriott. They discussed the Army of Occupation—the term President Truman began using after the occupation of Japan—and what Bill might do in the future. Mr. Marriott tried to be positive.

"Surely this war won't last much longer. If you change your mind, you'll always have a job here," he said. "I like your attitude, Boy. You can always figure out whatever I throw at you," he'd said.

Bill took his final pay and said his thank-yous before he told the kind man goodbye. The two shook hands, and Bill broke into a run to make it home in time to finish his chores. The only two cows left on the farm were the two standing beside their frozen troughs, shivering, waiting as the temperature fell. Their lives depended on Bill's ability to break through the ice and find water for them. He took his father's axe, chopped through the ice covering the pond, and filled his two buckets with water.

The cows' warm bodies rubbed against him as they ambled by, as though to say thank you, and he stroked their heads as they passed. Once they were settled in, he fed and watered the last of the chickens who had sheltered in the barn. When he took his axe and headed back to the wood pile to split logs, he felt tears coming. The day had finally gotten to him.

The sun went down long ago, and he was tired and hungry. He swung that axe at those frozen logs so wide and hard that ice chips flew with every stroke. It made him feel better. The snow began to fall. He stopped cutting logs long enough to catch a few flakes on his tongue. From the west he felt a little breeze, and to the east he could see the evergreens sway. Behind those trees lay the forest where he and his father sometimes hunted for deer, and that night a doe and her two yearlings stood still across the field, just in front of that forest. The little family watched him, quietly, looking thirsty. So he filled two more buckets of water for the deer and set them out on the far edge of the farm. The last of the birds going south flew overhead. He saw the flock disappear above the trees as the deer drank from the buckets, but the spell didn't last.

He wasn't a bird. He was a crew-cut kid wearing boots covered in frozen mud, taking firewood to his mother and thinking twice about chopping up another of God's trees. He piled a load of firewood into his arms and hurried on into the house.

The wind was picking up. His father was away, and he hoped his mother had left food on the table. She was probably cleaning up in the kitchen, absentmindedly, waiting for a something she could not define. Her challenge had always been living without her husband, but now, with Adele married and gone, Jack away working on a ferry boat, Mama Evelyn was lonely, even though she still had five children at home.

He piled the firewood inside the door and hurried into the kitchen, only to see that the table had already been cleared. He'd been so intent on food that he hadn't seen Jack at first, but there he sat, at the far end of the table reading a newspaper. Nobody told him Jack was coming. Jack was his father's favorite son, the high school hero and fullback for Sault Ste. Marie's high school's football team who had made an unforgettable 440-yard dash before he left home two years ago and joined the Coast Guard. Jack pushed his chair back and stood up to greet his brother.

Bill looked him over. He managed a ferry for the Coast Guard and wore the uniform with creased trousers, fitted jacket, absurd buttons, and all. His shoes were polished and his hair buzz cut. Altogether, thought Bill, Jack was still handsome but he looked taller and thinner. He knew Jack's ferry was docked in the next county over, which was why he rarely came home, much less bothered with his younger brother. Jack gave Bill a rare slap on the back.

"Got a change in location, Kiddo. They tell me I'm going to live closer to the locks now, and I'm glad of it. Gives me a chance to come home more. Now you tell me what's going on at the old school." Bill returned the greeting, then stepped back. Didn't he know?

"You're looking good, Jack. I know Mama's glad to see you."

With no food on the stove, Bill poured himself a glass of water, grabbed a slice of bread and was about to go upstairs to change out of his work clothes when his mother came into the room with Eva following in her footsteps. He showed his mother the firewood. She nodded at her tired son, then turned to take Jack by the arm and lead him out of the kitchen and into the living room where the two would fall into uninterruptible conversation.

On his way up the steps, Bill could only wonder whether Jack would come home to do the work that needed to be done around here after he, Bill, left for the services. He doubted it.

A little later that night, his father came in the kitchen and ended any hopes Bill might ever have had for finishing his senior year with his classmates. He never had a chance to ask for a reprieve. His father made it clear.

"Your mother and I will not change our minds. I've already registered you for the Army, so you need to sign these papers. You will report for duty on January the first, 1947."

Bill looked down at the papers and couldn't believe what he saw. The blanks were already filled in: height: 5'11", weight: 163, hobbies: fishing and building model planes. His cause was lost. He signed.

The Christmas party came and went, and the year ended far too fast. He was curious about so many things and had hoped to find answers by finishing school, but that hope was dashed.

The night after Jack left, Bill put on a clean shirt and his last pair of pants that held a crease, added a tie, and set out on the long, dark walk to Virginia's house. He needed someone to talk to, and he knew she would listen. As the house came in sight, he began to whistle, and the closer he got, the louder he whistled. He didn't want Mr. Buzzo to pull out his pistol before he had a chance to identify himself. He even sang the school alma mater as he walked up the steps to Virginia's house and knocked on the door. Her father was frowning when he opened it.

"What are you doing here at this time of night?" He let Bill come in from the cold but didn't give him a chance to say anything. "Virginia can't see you. She's already gone to bed."

But Virginia had heard the whistling and was already halfway down the stairs in her warm blue robe when Bill, afraid he was about to be booted out the door, blurted out to Mr. Buzzo that he was leaving the Soo the day after Christmas. Virginia reached the bottom of the stairs in time to hear, and with one more step looked up into Bill's eyes without saying a word. She just stood there beside him.

Bill didn't try to hold her hand because Mr. Buzzo was there. The two just stood and looked at each other while Mr. Buzzo talked. They weren't listening. Virginia's mother came in at about that time.

"Mr. Buzzo"—she always called him Mr. Buzzo—"this young man walked all the way here at this time of night and he probably has some important things to say to Virginia."

"That is not your concern, Mrs. Buzzo," said Mr. Buzzo.

"We can discuss that later, Mr. Buzzo. I think we should leave them alone for a few minutes and let these two young people have a chance to talk."

Bill wanted to tell the whole family at once that he was leaving for good, to join the Army, so he told them all, right there, before Mr. and Mrs. Buzzo left the hallway. When her parents left, the two just looked at each other.

"I'll write," Virginia said.

"I'll write you, too," said Bill.

Finally and with trepidation, Virginia asked, "Do you want me to wait for you?"

Afraid to say something wrong or something that might offend her, he simply said yes. He finally got up enough courage to hold her hand for a while. Then he asked her to tell her parents goodbye as he put his wet boots back on and walked out the door. It was quite late by now, and Bill had plenty of time along the way to think serious thoughts about what he just said and what he might have said or should have said, perhaps not what his parents wanted him to say, but because he didn't know what his parents wanted him to say, he resigned himself to the fact that he couldn't have said it.

At his house, those last weeks in December ended mostly in silence. He turned eighteen on Christmas Eve 1946 and finished packing. On January 1, 1947, his mother and his five younger siblings were all at home. Bill waited as long as he could for somebody to say something, but nobody did: neither his mother or his father, who left for work that morning, nor brother Mort, who slept in that day, or Adele, now married and gone, not even the dog.

Memories of having a hand to hold as a lighthouse keeper's child lifted him a little. But there were other times when the sea had roared around him in the storm, and his body had trembled. He closed his bag and went out the door by himself to walk the three lonely miles on frozen tracks to the station. He would wait there in the dark until the train to Illinois came along. Then, by morning, he would arrive in Waukegan and find his way to the Great Lakes Station where he would report for duty.

CHAPTER 3

The Blues

He was the first one aboard the train that night, but a few more young men joined him as the train made stops. By the end of the trip, half a dozen young men disembarked in Waukegan. They knew they were headed the same way and walked together to report for duty. Most of the boys at the Waukegan District station came from the heart of Lake County near Chicago, a long way from the Soo, but all of them would be assigned to the Fifth Army.

They stood tightly together in front of a sergeant who administered the oath of enlistment, then pledged their fealty to the Army and their allegiance, in unison, to the flag of the United States. Then and there and loudly, the official told them that every man standing before him was enlisted "for the duration of the War or five years or other emergency plus six months, subject to the discretion of the President or otherwise according to law."

A week passed before Bill and the others were done with the legal stuff. They were now waiting for the bus that would take them on to Ft. Sheridan, Illinois. The chatter was about war: missiles, two-way radios, radar, and computers. The bus came, and the sergeant in charge made them repeat the whole pledge they'd already made about the duration of the war before he let them on the bus.

The entire "for the duration" pledge had to be repeated every time a new batch of young men got on board. Whenever a new recruit arrived or said goodbye to his family, he would hear those words, and the more they heard, the more it worried them. What did it mean? All they could do was follow orders and wait.

Bill waited, and while he waited he began to understand what else was going on and why the War had not ended. The communists or rather the Soviet Union were still a real threat to all democratically elected governments across Europe. Bill watched the others in the Army of Occupation leave for Germany or France or parts unknown. He had no idea where they would ultimately send him. Waiting was the hardest part.

The twenty or so others on the train just like himself talked about where they came from and what they did or wanted to do. The talk was all "Navy" or "Army." Bill was mostly interested in the Army Air Force, its new engineering battalions and schools of engineering he'd heard about somewhere. He might have a chance at that. He saw himself flying Mustangs, P-51s or even F-51s or jets, but first he had to go through processing at Ft. Sheridan, Lake County, Illinois.

The bus let them off at Fort Sheridan close to nightfall. They were assigned to barracks where they tried to get some sleep. The next day, the sergeant woke them before daybreak and ordered them to form a line in front of Fort Sheridan's infirmary. That line was so long he could have wound it around a city block.

The tall, skinny guy in line in front of Bill turned around.

"Are we in the right line?"

"Is there another one?"

"There's gotta be. This one goes on forever. Bob Johnson—Hennipin, Illinois," he said and shook Bill's hand.

"Bill Brosco—the Soo."

"You mean the locks?"

"Not right on 'em, but near enough. I lived in a lighthouse near the locks when I was a kid, but we moved to Mobile. I'm back in Sault Ste. Marie, or I was."

"Yeah, we've got locks in Hennipin, too…but nothing like the Soo. We're on a tributary of the Illinois."

"Should be some good fishing."

"I know the place," said a deep voice behind Bill. "Hennepin! Not that far from Chicago. A little village, right? Fishing and hunting?" He pointed to himself. "Dave Beaversdorff. People call me Beavers."

As handshakes were shared, they knew they would be Bill's friends.

"Sounds like at least a fishing trip," said Bob.

"Yeah," said Beavers. "No tigers and lions!"

"Season for deer, but then by the time we get through this line, it might not be!"

The line moved a little more, and Private Lamont Rollison joined the conversation. The four of them moved on together.

In the great scheme of things, the Army shuffled recruits and privates alike into lines like these. Veteran pilots coming back from the War took precedence over the newbies for the infirmary, so the four had plenty of time to waste. They moved like snails.

Beavers wore a civilian, double-breasted suit just as he always did when he went to work, tie and all. The rest? Most dressed like country boys, but it was hard to tell. When Bill's turn came, he was just one more naked guy suffering through an Army medic's scrutiny.

"How many shots have you had, Soldier?"

Truthfully, he told the guy holding the needle, he hadn't had any.

By the time they finished with him he felt as though he'd tangled with a porcupine, but those shots weren't the end of it. He followed directions to the next line and the next and emerged with bandages on both arms as he entered the room where he had to take the Army's IQ test, which included questions like "show me your left ear." That one was designed to find out whether or not a man could tell left from right, a skill necessary to let the Army know whether or not a recruit could obey a command to shoulder his gun on the correct side.

Bill was sure he would breeze past the barber because Adolph always insisted his boys keep their hair military style; but Army barbers found lots more hair to cut, which Bill thought was unnecessary considering recruits had to wear helmets all the time. After the medic took Bill's medical history and finished all his punches and probes, he sent Bill off to yet another building to get his gear: one new Army-issued blanket and one towel plus clothes and boots that did not fit.

Bob looked around and commented he needed a bathroom. Had anybody seen one? They had not, but the master sergeant soon made it clear that as enlistees, they had to wait for him to tell them when and what they could do and when and where they could go. By the time the man finished, there wasn't a single soul in the pack who thought he could hold in any longer. Finally the sergeant finished his speech and marched them off together to quarters where they would share thirty-six back-to-back toilets. Bill returned to the grounds barely in time to hear the sergeant's order, "fall in," before he marched his recruits off to the mess hall.

The next day, the sergeant handed him a letter dated the day he left. It was from Virginia, and his spirits rose—it was the first thing he'd heard from anyone. He read it through, and although it was mostly about school and news about his family, he read it twice.

Virginia was true to her word and her letter included a carbon copy of the page she'd written for the yearbook. Under "legacies" she described in a few words the most important thing about each member of the Class of 1947, and Bill would forever appear as a "frustrated engineer." And right there, somewhere in a footnote, she had underlined the words "1947, a class without Bill Brosco."

The next week, after morning drills, the sergeant handed him another letter from Virginia, and this time not a very happy one. Brother Mort had been suspended from school and sister Rita was getting fat. Bill read all the way down to the bottom, but Virginia didn't ask about boot camp or how he was. She wrote things like "have fun with the boys." To her, he was just out with a bunch of the fellas, going from camp to camp for no real reason. The war was supposed to be over.

He stuffed the letter into his pocket before going on to yet another line to await the fatal cut. The men in line were waiting to receive their orders. Bill got his orders and moved on into line waiting to get in the barracks when he saw a curious thing. One master sergeant was moving about freely, back and forth from the men in the line to a smaller building next to headquarters. He was calling out names, and when a man answered, he would take him out of line and over to the building across the way. About one out of every eight or so men would go to that other building. He could hear mumbling up and down the line. Perhaps they'd done something wrong or had a record or something.

"Why them and not us?"

Beavers just laughed. "Those guys aren't crooks; they're special! They were all aviation cadets when they were in high school, and they're the ones the Army picks for special duty."

Beavers was older than they were, so of course he knew such things. Bill waited until the sergeant came to pick another man out of the formation and leaned in to hear what was said.

"The sarge just said something about a 'college training detachment.' I think that's what I heard." Bill put it out of his mind; not much use dwelling on it now.

As the day wore on, he made a mental note of how many were pulled out of line and concluded that only about half a dozen men had any college at all. He let that set in. He knew that he and Bob never finished high school, but he didn't know about Rollins or Beavers. All they knew about Beavers was that he had a wife back home; he never mentioned high school. It didn't matter. The Army had no rejects.

It had been one long day, and they had another break. They'd just come from the mess hall and were lined up in alphabetical order in front of a room full of long tables with men in uniform behind each table. The B table was right outside the door, and Bill saluted the man who was about to interview him. The man pulled out his record and looked straight at him.

"This is your last stop, Soldier. You still need to sign here to become part of the Fifth Army Air Force," and he turned the paper around so Bill could read it for himself.

"Sir, it says five years and I know I have to serve five years, but it also says I agree to go anywhere in the world the Army sends me. Sir, may I make a request? I've had some experience with planes—I even flew one back in the Soo. I want to be part of the Army Air Corps after Basic. Is that possible?"

The fellow looked back at his records.

"Says here you liked to build model airplanes," the guy grinned. "That what you flew? It says here you worked for Northern Electric. That so?"

"Yessir. But Nossir. Yessir, I did fly a Piper. My friend's dad let me fly his Piper."

"It don't matter. You're pretty much all going the same place: Basic Training Wing to the Indoctrination Division." He wrote something on Bill's record, stamped it, then looked up at Bill.

"Lackland Air Force Base... that's the only place the AAF uses for basic training, but that don't necessarily mean you're gonna fly. You need a college degree to fly missions in this war."

The man dismissed him. He had nothing else to say. Bill told Beavers and Beavers told Bob and it turned out they were still all going to the same place! Beavers didn't care much which branch of service they sent him to. Basic training was basic training. All he wanted was the branch that paid him the most.

It was a long summer. In the weeks ahead, they did get assignments—mostly maintenance tasks for the base. In the end, the men concluded that no matter what they asked for or wrote on some official document, their wishes were not worth the ink they used up. Their superiors could see, they said, but probably would never read those requests, and they definitely wouldn't take time to hear them out, all of which was mostly true. A man's only consolation was not being the only one, but they would never give up on trying.

Bill's best ray of hope for flying came when the sergeant called them into formation and announced that President Truman had just created a "Department of the Air Force." With that, the sergeant saluted his men and dismissed them. Nothing changed except that Bill and Bob doubled their efforts to find out what their fate would be.

By September, they'd given up. But the day finally came when the captain counted off his men and arranged them in alphabetical order to march to the auditorium. Bill, Bob, Beavers, Rollinson and the rest of the now Privates First Class were going to Fort Warren in Wyoming. Bill looked at his new assignment: electrician. Not exactly what he had hoped for, but he read further and saw he was being assigned to "Fort Warren, Wyoming, School of Engineering, Refrigeration Unit." He was happy to study engineering, but nothing was said about flying. Still, School of Engineering? Not bad.

The three Bs and Rollison were at least headed for the same place together again, but they were all assigned to different units. They worked themselves into various states of fitness; Basic Training was about to end when the real miracle happened. On September 26, 1947, Chief of Staff for the Army, Dwight D. Eisenhower, announced on the radio that under the President's new Department, he had separated the Army Air Force from the Army, and the AAF was henceforth the "United States Air Force"!

Bill and the other wanna-be pilots jumped up and down when they heard it. There was going to be a ceremony at the Base, and all the men—no exceptions this time—would transfer to the newly created United States Air Force. Ike was a Texas guy, so the Texans celebrated for days.

Bill could not have been happier. Being part of a stand-alone Air Force and no longer the most minute cog in the much larger wheel of an all-encompassing Army meant his chances of flying were better, much better. Furthermore, this remote base in the comparative wilderness of Cheyenne, Wyoming—Fort Francis E. Warren—was about to be his home for a long time now, and he might just consider making his home out west, where there wasn't so much civilization. He couldn't possibly go wrong.

He had barely adjusted to Ft. Warren when the troops received disappointing news: the Training Center in Spokane, the flight school where Ft. Warren's pilots always trained, had closed. The four men spent their days once again doing push-ups and going on long marches and scurrying over obstacles.

The Air Force did not forget them. The new USAF reviewed each man's record after the transition. Someone noticed that Bill had worked for Mr. Marriott, and now that the troops had settled in and toughened up, the sergeant called another briefing. As each man presented himself, the sergeant handed him his new designation. Bill's was "Private First Class Electrician." He had to ask what that meant. He found out all right: he was now not only responsible for refrigeration but for all things electrical including repairs and wiring and everything even remotely having anything to do with electricity anywhere on the base including teaching new recruits.

William R. Brosco

By that time, Uncle Sam had closed several air bases; pilots were coming back from the war by the hundreds if not thousands; and the government didn't know what to do with them. Service men were on hold, and Bill's hopes turned to resignation. For all he'd gone through, the Air Force had abandoned him to a civilian teaching job at Fort Warren, and he had three and a half more years to go just to fulfill his five-years' duty with no mention of further college much less a future that included piloting. The lieutenant's last briefing ended with the now familiar "for the duration of the War or other emergency plus six…" which all soldiers now mouthed dejectedly along with the sergeant or major or whoever was doing the speaking. Whatever joy they might hope to command on this base would inevitably depend upon how that same sergeant or major interpreted those words "other emergency" as they applied to each person. From then on, the men tuned into the news regularly and read every paper they could get their hands on, hoping they could figure out what would happen next.

By the time 1948 ended, Bill was a sergeant headed for Lackland Air Force Base in San Antonio, Texas. Lackland had once been the AAF Training Center and was now the USAF Training Center and the only remaining basic training wing for the Indoctrination Division, the unit that taught a soldier to conform.

It was also the base where men earned their wings. For the first time, Bill figured he might have a future in the air. It was time to ask for time off to go home and see if Virginia would marry him. Meanwhile, he resigned himself to even more push-ups in the Texas heat, hurdles in the field, and avoiding men who joined the services just to kill something or somebody. They all belonged to the "Fighting Fifth," and to some the "fighting" part took precedence. Most of the PFCs were damn decent, though.

The Fifth Army Air Force was meant to fly, but Bill still wasn't flying. He was making repairs on heavy bombers and grounded aircraft, repairing the flotsam and jetsam left over from the last war when the last of the occupation forces left Europe and Japan. The Fifth was just one of the many countries' "Occupation Forces" stationed all over Europe, overseeing reparations and preventing more wars.

It was then, in 1949, that President Truman began gathering America's allies together to counter threats like those from the Soviet Union, who had made it very clear they planned to take over smaller nations, especially those in Eastern Europe. These allied countries joined forces to sign the North Atlantic Treaty, the international agreement from which the North Atlantic Treaty Organization (NATO) grew. All the countries that signed that treaty promised that a threat on one nation would be treated as a threat to all. Member nations had the right to use force, even nuclear force, if necessary to stop it.

Like so many, Bill had no idea how or when this cold war might play out, so he spent time on the shooting range earning qualifications and, because he was still deaf in his left ear, agonized over not being able to fly. To him, being part of NATO meant always being prepared.

About that time, he developed a horrible rash all over his body, but it didn't really upset him because a rash wasn't likely to keep him from flying, but his damaged left ear could. According to the base doctor, there wasn't anything he could do about either one of them.

Then came a minor miracle. In the process of picking up his latest gear, he learned that each airman had to wear a very specific kind of ear muff—one with a built-in radio receiver—when he went flying. The miracle happened when Bill tried on his muffs and could hear just fine! He couldn't wait to tell the Base doc about it, and the doc gave him the thumbs up and assured him that when the Air Force began training pilots again, Bill had a real chance of becoming one of them.

But he still had a rash when he weaseled his way into pilot training and after he flew solo and qualified for his wings. Just being qualified was not enough for Bill; though. He wanted to fly jets, and the only way he could see to do it was, first, to earn his high school diploma by mail and then go on for a college degree. But first, he had the grand Wings Ceremony to look forward to. No soldier who'd earned his wings would ever miss that.

It seemed like years since he'd been home, and he truly missed his family. He would invite them all to see him get his wings, including—and especially—Virginia.

"My family's coming!" said Bob when he got his letter. "Have you heard anything?"

"Not yet. You know, her letters come slow, but when they do, she always says it's time for us to have some time together. She'll come."

"Yup," said Bob, using his best country talk. "Ain't nobody don't want to see a man get his wings."

But when Virginia's letter arrived accepting his invitation and telling him she was making plans, he developed another terrible rash, this time on his legs. By the next day the rash had spread to his chest and face. He was terrified that she, who had never seen him without his hair combed and his shirt freshly pressed and with a crease in his trousers, would see his disfigurement and he would become her pariah. Beavers came around and did the moon thing so they could all see his rash. It was worse than Bill's.

That rash was the absolute last thing he wanted Virginia to see. Making another trip to the infirmary and getting naked was not a pleasant thought, but he forced himself. The doctor took one look at him and shook his head, then gave him chamomile to rub on his body and sent him back to his barracks. Weeks later, with the rash growing worse, Bill went back to the infirmary.

"Ok, Doc, I've got to know. What is it?"

"Well, I can't say for sure by just looking, and I don't have any way to culture it, but it looks like ringworm to me. Whatever it is, the whole camp has it, and now I'm treating it with Whitfield's."

Later that night, Bill spread the oily stuff all over his body, and he and Bob and the others sat around all greased up after taps making jokes and planning for the ceremony. Rumor had it that they would have beer afterward. It had been a long time without.

Bill's rash slowly improved, so because he hadn't heard anything from his family, he decided to call them, to encourage his mother and father to come and bring all his brothers and sisters, if they would. Maybe they wanted to come, too.

He especially wanted to see Adele. Big Sister Adele had always stood up for him, but she wasn't home the time his mother hit him so hard. He told her later. He told her everything, and she told him about everything going on in the family. He had just received her letter telling him how happy she was to hear about the muffs and was so proud to hear that those muffs enabled him to earn his wings. She wouldn't be able to come, though; she had baby Dave and couldn't get away.

Virginia thought her parents were going to let her come, at least that's what she said in her last letter. He told Bob he couldn't wait to be with her.

Bob had been standing across the room, listening, but when Bill mentioned he thought Virginia was coming, Bob stepped across the room and leaned in closer.

"Turn your head up." Bill looked up at his best buddy. "What's that on your face?"

"I don't know. My body looks better. The doc just said treat it with this stuff— Whitfield's. So I guess it will work on my face, too."

"Well, I hope so. It's pretty bad. You might oughta tell Virginia."

There was no way Bill was going to tell her. In spite of rumors around the base that changes were in the wind for the base that might affect the ceremony, he and Bob made plans as though nothing would ever change, and that made the weeks fly by.

By the week of the ceremony the Whitfield's had lessened the rash, but Bill still hadn't heard anything from any of his family, and nothing more from Virginia since she accepted his invitation. But he still had hope.

The day arrived, a beautiful day in May for the men at Lackland AFB to gather on the field with time to enjoy each other's company while they waited for their Captain. There would be no marches in Texas that day or fitness exercises before lunch, just a debriefing of the protocol required during the actual event and afterward at the reception. The mess prepared steak and potatoes for a crowd, and the new pilots left their barracks in good spirits, looking sharp in their new blues as they formed ranks and marched into the auditorium, eyes front.

When they were all seated, the captain, now Master of Ceremonies, began calling them out one at a time to come forward to receive their wings. As each man stood before him at attention, the captain gave a little prepared speech— something about each man that, like legacies in the yearbook, made the others laugh. When Bill's turn came, the listening audience learned that Bill's reputation as a gymnast served him well.

"You're soon going to see Bill Brosco repairing part of a plane while doing a hand stand" brought laughter. Then the captain called Bob up and gave him a new name.

"This here's Bob Johnson. We call him Slinky, which has nothing to do with him being tall and skinny or falling down the stairs. We call him Slinky because he's acquired a reputation for robbing the kitchen after hours!"

Bill never really heard what they called Beavers and Rollison because he kept turning to look at the door as the last of the latecomers straggled in from the rear of the auditorium. There was not even one Brosco. And there was no Virginia. Bob told Bill not to worry; he would pin Bill's wings on, and that was what happened.

Later, husbands kissed wives and fathers held babies in little family groups. Soldiers and families paraded together to the mess hall for food and beer. Bob and his parents invited Bill to sit with them and did everything they could think of to include him, so Bill spent most of the rest of that very long day with Bob and his family.

Reveille played without a Brosco or a Virginia in sight. All guests departed.

Bob returned to the barracks to find Bill sitting alone, his chin way down on his chest and his shoulders slumped. If he were a man to cry, he told Bob, he would have cried, but he wasn't. And perhaps it was just as well that Virginia didn't see his rash.

"Just hang on," Bob said. "I'll be back in a minute."

It wasn't long before Bob returned with two pints of rapidly-melting ice cream and two spoons. He gave Bill one, and the two of them sat on the bunk side-by-side, eating their pints until the ice cream was all gone. After that, they both knew they would stand up for each other whenever they could, whenever there were hard times.

They left Lackland and returned to Fort Warren, Wyoming. In the days that followed, each of the four friends received the World War II Victory Medal, and Bill qualified as a sharpshooter, which didn't surprise him at all after his years in the Soo. All that busyness took his mind off things … until he received not one but two letters from Virginia.

The first one said how sorry she was that she couldn't make it because in the end Mr. and Mrs. Buzzo wouldn't let her go. They thought her too young to be hanging around an air base on her own no matter how hard she begged, and they declined to go with her. Then there was the second letter: unlike the off-putting tone of the first, this letter was an invitation. Her parents wanted him to stay with them over Christmas. She wrote that they wanted to congratulate him on his wings in person. Bill read the letter as a kind gesture, one that could turn his blues if not sunny at least less blue. But none of them had earned his forgiveness, not yet. Nevertheless, he wrote back that he would ask for leave. He would take whatever he could get.

When the Air Force assigned him to a worn-out old T-54 prop transport plane that had been riddled with bullet holes in the War, his excitement exceeded his good sense. The word aviator made him smile, at least a little. But he still had the rash.

On his next trip to the camp's doctor, Bill figured out the real reason he still had the rash. Why hadn't he realized? The doctor had said he was sorry he couldn't fix the problem, and he'd laid out reasons like he had no way to test anything. and it must be something new because so many of the troops had it, and other excuses. He'd even asked Bill if he'd been "going out with girls," but Bill assured him he had not and would not.

"I know where they've been, Sir," Bill told him. He told the same thing to anyone who asked.

The doctor shrugged his shoulders and finally admitted he was not equipped. He'd come to the base as the director of the Visual Acuity Clinic and was only supposed to perform eye exams. He apologized, but he said he couldn't help it if he was the only doctor in town.

It was time to ask for leave to go home for medical treatment and more importantly, to marry Virginia. That is, if she would marry a man with an itch.

He thought about Mrs. Buzzo's invitation. He supposed he could spend

Christmas wherever he wanted, but her house seemed logical, especially with all the turmoil going on at home. She lived close enough to his parents' house so that he could visit his father and mother without having to stay with them, without having to be around all those younger brothers and sisters who seemed, according to their letters, to be fighting all the time. But first, if he wanted to spend time with Virginia, he would have to ask for the moon: he had to ask for extra leave.

The Captain first suggested that he might stretch a medical leave, should he need time for a wedding, and he'd do what he could. He'd call Bill later.

The day Bill checked back, the captain told him he had found a solution: an honorable discharge—not a leave. This leave, he told Bill, would be flexible, but it had a contingency: Bill had to re-enlist afterward, and the Air Force would say when. They shook on it.

The last letter he'd received from his mother Evelyn was actually a report, and definitely a bad one. Eva was terribly sick, and Mort had been arrested and was in jail. She didn't say what happened, but Mort was only sixteen years old, and Evelyn told Bill not to worry about it until he got there. There was much that needed doing.

As he packed for his first trip back to the Sault, he took a good look at his choice of boots, flight jackets, caps, T-shirts, and dress blues—all in good order, but not likely to be exactly what he would need when he was visiting the Buzzos. His civilian clothes had not budged from the bottom of his trunk for way more than a year, and when he dug out his white shirt, he hardly recognized it. The barracks didn't provide ironing boards, but surely Mrs. Beavers would have one. Before he knew it, he was knocking on Beavers's door, feeling foolish indeed. Just as Bob had done before, Beavers told Bill to look up. Bill's rash was better but not gone.

"Dear God, Bill, I'm not surprised! Didn't you know you've been going to an eye doctor? He's not even a regular doctor!" Bill told him he already knew.

Mrs. Beavers just shook her head. Of course he could use her ironing board. So that afternoon he left with shirts and trousers pressed and folded, and with a head full of advice from the Beavers family, most importantly he was to take flowers to the Buzzos. He promised.

The last thing he did before he left Wyoming was to ask the sergeant if he might request dependents' housing upon his return. The sergeant told him to forget it—there wasn't any available housing and there was a long wait list thanks to all the returning soldiers. Bill's hopes fell, but he wrote a letter to the head of the squadron anyway, saying he understood the shortage and begged them to put him on a waiting list. He would keep on asking. He knew Virginia would love living in Wyoming.

If he felt foolish asking for that ironing board, he felt even more foolish carrying a huge bouquet of flowers on the plane to Michigan, but somehow those flowers symbolized commitment. At least he hoped they did. He knew he had changed and so, probably, had she. He would somehow manage the flowers on the flight. Indeed, the stewardess agreed to take care of them in the cockpit. So the bouquet flew with him and rode with him in the taxi. He had to fight the bouquet every time the taxi took a sharp turn, but both taxi and flowers made it safely to the Buzzo home.

Her parents were there when Virginia put them on the coffee table, where they continued to bloom the whole time the two were together. Somehow, in a room full of flowers and in a rare moment with no parents present, the two managed to become engaged.

Bill could now no longer avoid going home; he had to tell his parents. The entire Buzzo family went along with him to announce upcoming marriage plans for November. Adolph didn't object, so they set the date for the 26th. Bill would be twenty years old; two years of his five-year commitment in the service would be up. He told them all that he and Virginia would start life together at Fort Warren, but as soon as he became a civilian, they would move somewhere else. He said he wanted to marry Virginia more than anything and had already requested family quarters. He promised he would make sure Virginia would have a house, and the

Buzzos left. Then, as soon as the Buzzos were gone, he asked his mother about Mort.

"He wanted to borrow his father's car, but with his track record, Adolph refused him and accused him of doing nothing but ask, ask, ask. He even told Mort to get out; he was tired of hearing it." She paused until Bill encouraged her to go on.

"He ran away. We didn't hear a thing from him for about a week … " She swallowed. "Then one night we heard sawing and banging on the garage door and got up just in time to see him drive off in your father's brand new car. The next thing we knew, a policeman was knocking on the door. They found your father's car in a ravine. It seems Mort drove it, and when he'd finished with it, he pushed it off a cliff. He was angry at the world and thought nobody would know, but somebody saw him do it and called the police. The police arrested him. It was only weeks later that they found out whose car it was. In the end, after your father heard the story, he pressed charges. He's still in jail." All of a sudden Bill saw his mother looking helpless and vulnerable.

He put his arm around his mother's shoulders. "Don't worry, Mama. I'll see what I can do while I'm here."

"Will you go see him?" She had tears in her eyes. He had to tell her he didn't have that much time, and the jail was in another state.

"I don't know. The Air Force gave me an honorable discharge just long enough to come home and marry Virginia, but I have to re-enlist immediately afterward. I have no choice. I agreed to it and don't think they'll give me much time."

On November 29, 1949, Miss Mary Virginia Buzzo and Mr. William Roland Brosco married in Mt. Clemens, Michigan. According to the papers, the bride finished high school and the groom was a Sergeant Career Instructor in the Air Force. What the groom read between those lines was "this Sergeant has to get his degree, and soon."

Now that he had his wings, and as soon as he requested permission to live off base, he applied to take classes at the Engineering School in Colorado. He expected an answer by the time he got back, but as things stood, he would have to stretch the $200 a month he received as "sergeant, married man" as far as it could go in Fort Warren, Wyoming. He had to tell his mother that he could no longer send her his check.

Back at the Base, he forged ahead believing he was the luckiest man he knew, even without a degree. Not only did he have a wife, but he was one of very few whose request for married family living quarters at Fort Warren had been granted. The day he married Virginia, he received his re-enlistment orders as he'd been promised. He would report for duty at Fort Warren as soon as he could get a flight out.

The first thing the couple did when they got to Fort Warren was to go look at their new apartment—a place of their own in an institutional-looking big brick building in Cheyenne. He could hardly believe it and neither could Virginia, but no sooner had they walked in those doors than she was already hanging curtains and learning to cook on a Kelvinator.

When he reported for duty at the Francis E. Warren Air Force Base, he discovered he was to be not only a Powerman but an instructor in engineering. He was one of about a dozen of Ft. Warren's engineering instructors, and all the instructors were given an opportunity to earn an engineering degree through Colorado State. All they had to do was catch a bus four nights a week, attend a few lectures and classes, and catch the same bus back to the base in time to study.

He left the duty officer's quarters disbelieving. Was he, Bill, going to be the recipient of good fortune? He would never count on fortune, but while it was smiling, he would take advantage of every bit of it he could get.

For the first time in his life, he had a real home, a place where he was welcome, and there was this incredible chance that he might earn a degree. It would mean late nights and hard work, but that was something he was used to. He would stay right there as long as he could, study hard, do double duty as both Powerman and instructor of engineering for the Base ... until they sent him somewhere else.

CHAPTER 4

Peace That Is No Peace

In all history, this is the first time that an allied headquarters has been
set up in peace, to preserve the peace, not to wage war.
— General Dwight D. Eisenhower,
Supreme Allied Commander of NATO

F all 1950. The USAF had, ever since the USSR had armed themselves with nuclear weapons, planned for and developed strategic deterrents that could reach the Soviet Union in a day. But so did the USSR. Each owned intercontinental ballistic missiles (ICBMs) that could reach the other in minutes. Western Allies and NATO countries that had been divided between the two superpowers were always on the alert.

Sergeant Smith was handing out "Good Conduct Medals" to the troops at Fort Francis E. Warren AFB and sending them on their way when they heard that the Soviets had bribed North Korea to attack Kimpo Airfield in South Korea. The United Nations' Security Council had directed the Fifth Air Force to evacuate U.S. citizens from Korea.

In spite of the alert, Bill picked up his orders and saw he was going to Eglin Air Force Base outside of Pensacola, Florida. He could not believe what he was seeing. He'd been there once before, with Beavers, and it was his favorite place on earth. Virginia would have to stay in Wyoming long enough for Bill to get settled at their next base. He felt bad about having to leave his young bride; they had talked about such a thing and promised they would live together no matter what. It would be hard to be gone, even if only for a few months until he got settled. He hadn't been in the service that long, but he'd seen servicemen's marriages break up when wives were left behind and husbands were called elsewhere.

"I'll resign as soon as I turn twenty-one, as soon as I've fulfilled my duties for my country—I will always do that. But I can promise you there'll come a time when Uncle Sam doesn't need me anymore!" Virginia believed him; she even encouraged him.

"I know. I don't want you to ever do something you shouldn't, but I also know that you'll always have a job back in Michigan with Mr. Marriott, and Michigan is my home, too. It's a long time before you'll be twenty-one, and I don't want to be here by myself, but I also don't want you to make any quick decisions for my sake." It had been a tender moment.

The reality, which neither one of them ever thought of as a reality, was that he'd signed up for five years' active duty, and he'd barely served two. World War II had gotten old. Surely it wouldn't be much longer before Truman declared World War II over, but the only thing Truman had done so far was promise the world that the United States would always support democracies whenever or wherever authoritarians threatened them. The President made his promise official, and it went down in history as the Truman Doctrine.

Most people understood that those simple words in the Doctrine referred, most urgently and rather specifically, to countries being threatened by Moscow, especially by nuclear missiles. The world thought that only military dominance could prevent such a war, but that always involved secrets and spies and killing off vulnerable people. All of it seemed far away to Bill and Virginia. Bill's only thought now was to do everything he could to bring Virginia to Florida to be with him. She would love it, he thought.

The U.S. Air Force had called Bill to the "Gulf Coast Training Region" once before, for special training, but they didn't keep him long. Eglin was just a stopping-off place, a training station for a few airmen in some specialty before they were sent somewhere else. This time, though, Bill's stint at Eglin would be far more than a short visit. He had to bring Virginia. He wanted her to experience the Gulf Coast, where the waters were clear and the sun shone best.

He'd met a fellow named Dewey Destin there, and Dewey took him fishing for snapper and redfish in the blue-green waters of the Gulf of Mexico, fish that didn't grow in the cold waters of Lake Superior. He'd loved sharing a beer with Dewey and his crew afterward. He didn't even mind when they called him "Yank." The term never sounded bad in Florida.

This time, the USAF was staffing its new Aviation Division. Bill was still Eglin Air Force Base's Powerman, the guy responsible for testing electrical equipment, for overloading it on purpose to see how much it could stand, then regulating the tested system so it wouldn't explode or blow up a circuit. The Air Force always came up with something new to test or something old to improve, and there was always that risk of fire or explosion. With a family to look after now, Bill had to be more careful.

He wrote his mother and father back in the Soo to tell them his new job was going to be training enlistees to operate and maintain Eglin Air Force Base's mighty radar system. He was in charge of the four huge Diesel engines that ran the radar equipment all the way from Destin to the Louisiana border and slightly beyond. He told them he was working in a "climatic hangar," which was more or less a huge Quonset hut that had been turned into a place where everything electrical could be tested in subzero temperatures, but he couldn't tell them everything. He didn't say anything about more or less secret things like the JB-2, which was Ford Motor Company's copy of a German V-2, a surface-to-surface, pilotless flying bomb like the one Germany used in World War II, but hadn't been used since.

His family probably knew about the V-1 and V-2, because the Germans had set up bases in Poland to test them earlier in the war, not long before the German scientists who worked on them surrendered to American soldiers. Those scientists brought those rockets with them to the States, and he'd heard those rockets were in New Mexico.

He signed his letter, "Your son, Bill," then turned his attention to moving his wife to the Gulf Coast. It was time for a long overdue honeymoon.

After a long day among the radar stations and a delicious dinner at Beavers' home, it was time to listen to the radio. Beavers always turned it on after dinner, and Bill was almost always there.

President Truman had just come from a meeting with the North Atlantic Alliance and was about to address the nation. Hopes ran high that something positive would come of this meeting—if not the end of the war then at least something that would keep the Soviets from dropping its nuclear bomb on them. The Soviet Union had tested its last nuclear bomb as a threat to the West, and the world knew it. The North Atlantic Alliance, which included thirty European and two North American nations, had been working together ever since.

They had high hopes for the new year, 1950, as they tuned in to hear what Truman had to say. The familiar Missouri drawl come across the airwaves, and the news was good for a change. All thirty-two countries had finally signed the treaty, and all thirty-two committed themselves to protecting each other through NATO.

"War Eagle!" said Bob.

"You bet! Finally," said Beavers.

"It's done," said Bill, who grabbed a ruled notebook and began writing.

"What's that you're doing?" asked Beavers.

"What do you think?" Bill was scribbling out his first request to be relieved of duty. He hadn't told the others yet, but Virginia was expecting a baby.

Early the next morning he donned his uniform and reported to the master sergeant for the discussion he knew he had to have. Now that the war was just about over and now that they were a part of NATO, and because the President was treating the scrap in Korea as no more than a "police action," perhaps especially in light of the fact that Virginia was expecting a child, the USAF might consider a permanent honorable discharge for him so he could get a civilian job and be with his wife.

At the meeting with the master sergeant, Bill only got as far as saying something about the fact that the United States's responsibility to South Korea probably meant they wouldn't need him ... and since he couldn't fly jets ... when the master sergeant stopped him and began to brief him on the situation in Korea. Truman wasn't about to release any active-duty servicemen, he said.

Bill's excitement was short-lived. He told the master sergeant that he understood and that he knew there were new troubles brewing overseas, but the man continued. Money and military equipment had been pouring into communist North Korea—the Republic of Korea—from both the USSR and China for some time now. This meant the USAF and all other branches of the service were on high alert. North Korea was about to invade South Korea, and there were many Americans stationed there.

"Great Britain and the United States are responsible for all of them as well as all of South Korea. I just wanted to let you know that we may have to evacuate those Americans, and we're going to need pilots."

Did the Soviets think they could just take over South Korea without raising the hackles of Western Europe or the Americans? Well, hackles had been raised, and Americans would not abandon their ally. They never abandoned East Berlin when East Berliners were starving to death under Stalin's rule. America airlifted food to them as far back as '49.

What was the rest of Europe thinking? Like the United States, they were probably looking at their maps again to see how far Russian missiles could fly.

It was 1950, and according to the Truman Doctrine, the Allies had promised to protect not only each other's borders but also the borders of any democratic sovereign nation. All the United Nations, which included the USSR, were so committed. All the allies were supposed to protect the borders of Korea, both south and north of the 38th Parallel. The problem was, the USSR was responsible only for North Korea but wanted them both.

The master sergeant interrupted: he'd received a telegram for the Base Commander earlier and read it out loud to Bill: "All active-duty servicemen (and Bill was one) have their duties extended for another year, and anyone in this category at Eglin Air Force Base is to report for active duty in Korea immediately." Bill reminded his superior that he was responsible for Eglin's radar, and perhaps he should be exempt. The master sergeant read the announcement again.

"Sir?" said Bill, disbelieving.

"You are an active-duty serviceman and I see you've got flight training," he said. "The Air Force wants you to fly in Korea. It's now a combat zone."

"But Sir, I'm a sergeant, a non-commissioned officer. I do not have the college degree, and I understand a college degree is required for anyone to fly in Korea. If possible, Sir, please send me back to Wyoming to be with my wife."

"The country needs you—and for what you'll be doing, it doesn't matter where they have you stationed. You will have to fly. It looks like you're not only going to fly but you're going to be building runways for those planes! Don't worry, we're going to make whatever changes to your orders that might be necessary to make that happen!"

The master sergeant pointed at Bill's left shoulder and looked hard at him, then leaned forward and put his elbows on his desk as though he were sharing a confidence, which he probably was. Everything about this Cold War was a secret. Secret clearances were being handed out like candy.

"We're planning to put stripes on you so you can take this assignment. You will, of course, be a Tech Sergeant when you leave here. But if you go through our advanced aviation program for officers—just enough training to qualify you for overseas duty in Korea—you will arrive in Korea as a lieutenant."

"Sir?" Bill couldn't believe what he was hearing.

"Yes, Boy, lieutenant. A temporary one of course, because you really do not have a college degree. However, for the …" and he raised his voice a decibel … "duration of your service in Korea, you will be Lieutenant William R. Brosco flying for the United States Air Force!"

The more the master sergeant talked, the more urgent he sounded and the harder Bill was thinking. Maybe, just maybe, he might fly jets! They did have jets in Korea.

Then his boss straightened up as though he were reading Bill's mind.

"You don't have to worry about flying a jet. They are sending jet fighters to Korea, but most of the airfields that can handle jets are in enemy hands. We need planes that are not jets. The runways we have left in South Korea are too short and too rough for jets to take off and land. You'll train here on the Mustang. We need you to pilot a P-51 or F-51 Mustang, not jets. The runways South Korea still has are not only too short for jets, they're a mess. We need men to rebuild as many runways as they can." Bill's heart sank even further.

The man quit staring at Bill and looked back at his paperwork. "I see you're already familiar with F-51s. And with your work in cold weather hangars, you've got a head start in a country as cold as Korea."

Bill couldn't bring himself to tell the man he'd probably already forgotten everything he knew, because he never thought he'd need it. He just wanted to resign, but the master sergeant kept going.

"Maybe an F-6?" He twirled the paper back around to Bill. "If you agree, you'll finish the rest of your training right here at Eglin—after Lackland," whereupon he looked up at Bill with a huge grin and said, "and you'll begin collecting higher pay starting today! Take as long as you like to think about it, but I have to ask you to wait here until you decide."

"Would I have to leave right away?"

"No. You still have to qualify on the equipment, but the minute you agree to take this job, I'll see to it that we deposit your full benefits immediately."

Bill had never heard of such a thing as a "temporary lieutenant." Still, he was married, and if he re-upped, Virginia could have all the family housing allowances and other perks she wanted. It would make her life easier. And his, too. Maybe he could quit worrying so much about her. Then came the epiphany. He actually had a wife who ought to be involved in these decisions. He had to explain it all to her and asked for a minute to make the call.

In one short phone call, they agreed. She told him he could resign after he returned, but meanwhile he would have a job flying! And wasn't that what he'd always wanted? Furthermore, he would even have higher pay, and jobs were scarcer than ever in the United States thanks to all the veterans coming home from the big war.

The master sergeant stamped the paper authorizing Bill's new assignment. Then, just before he sent him off, he handed Bill a letter from his mother, which Bill stuck it in his pocket to read later. He saluted the officer, then left for the barracks. After supper, he eased it open.

"Dear Son, your father is well and still active in the Coast Guard. We haven't heard from Mort in a while. I'm glad you and Virginia are going to be together. I'm glad you have a good job, but you know that dealing with things like radar may be playing with God. You might want to see if they will assign you somewhere else. Your own Mother."

Bill folded it up again and stuck it in his flight jacket. His mother was French Canadian; what did she know about his job? What did she know about Korea or that he might be called up, for that matter. He would answer her, and he would explain the Korea situation and try not to alarm her by mentioning the treaty that was supposed to prevent war and that he didn't think North Korea had a nuclear bomb. He did tell her that the United States was obliged to protect South Korea seeing as how they had been partly responsible for dividing the country in two.

He followed the master sergeant's instructions and headed for the building that housed an office that would send him on to another building that was manned by an official of a wing of the San Antonio Aviation Cadet Center (SAACC), who told him he would once again leave for Lackland AFB in Texas to qualify on the only planes left in the States—old ones left over from World War II.

The U.S. Air Force had already informed all of the Fifth that Truman would need them. Indeed, it wasn't long before Truman called not only all reservists but also all officers who had been combat pilots in WWII back into service. For once, Bill was glad he wasn't a combat pilot. He believed his job at Eglin included enough vital defense responsibilities that the USAF would keep him in Florida to serve, with Virginia. Until he could arrange that, Virginia would be safe in Wyoming. She was already moving furniture around to make space for a baby crib in one corner of that house.

By August 1950, when Bill was twenty years old, the war—or police action—was officially on. Bill was no longer the engineer overseeing the ground power section at Eglin, but a married man going back to Lackland to become a temporary lieutenant while his new wife waited for their baby alone in Wyoming.

Texas was waiting for him when he arrived, and he hardly had time for lunch before he began his weeks of training without Virginia. They had indeed given him a relic of a WWII trainer to practice on, but it wasn't that different from what he was used to. In fact, it handled more like the old Piper he once flew in the Soo. Flying came easy to Bill, as many things did.

Bob and Beavers were there, too, all of them watching the newbies as they signed in.

"Seventeen and eighteen-year-olds! Where are the old guys like us?"

"They want young ones. They figure the younger they are, the more okay they'll be knowing they might have to die."

"I don't believe that."

"Well, that's what the thinking is anyway."

Later that week, Beavers brought in the latest gossip. A clandestine shipment of serviceable F-51s was already on its way to South Korea, so no doubt that's what they were going to fly. But then, no one really knew anything officially. All they knew for sure was that the USS Boxer had F-51s on board and had left from Alameda, California, destination unknown.

San Antonio finished with them, but the Air Force had not finished moving them around. They were all back to Eglin. They had more work to do, but when they looked for Rollison, he was gone. They asked around, but Rollison had left, whether or not for Korea nobody knew.

In spite of his now being a lieutenant, Bill found himself back in his old Powerman job, but working on a better radar system designed by the Massachusetts Institute of Technology and put into use overseas by the National Oceanic and Atmospheric Administration.

By the time Virginia came to visit in Florida, Truman had dissolved the Air Transport Command and had begun moving personnel with cargo and aircraft, preparing the troops for a war zone. Truman also extended active duty status for everyone, including Bill: nine more months of active duty.

That summer of 1950, the words Cold War took on a new and far more ominous meaning. The last day at Eglin came and went as though they'd only been there a few minutes. Virginia returned to Wyoming, and Bill stuffed his high neck sweater and arctic overshoes and muffler into his duffle bag. He counted. He had two towels this time and two wool blankets! Reality set in every time he put a cold-weather item in his bag.

That fall, long before they felt ready, they landed in Korea. There were F-51s on the tarmac, but those planes were not for either Bill or Bob. They had yet to see the planes they would fly.

"It feels like I've just been shot from a cannon," said Bill.

"More like landing with an emergency parachute," said Bob.

Before he knew it, they had all deplaned and were huddled silently together on the tarmac trying to catch their breaths. Bob needed that minute to gather his wits before he told the others that he actually missed everything he'd left behind and wasn't feeling too good about what lay ahead. Outdated planes and inoperative airstrips didn't help.

Bill and Bob reported to the Continental Air Command, which had been renamed the Tactical Air Command. The newer planes were assigned a pilot, a navigator, a waste gunner, and tail and side gunners, but Bill's plane was not one of them. Instead, it was a one-man wonder with heavy photographic equipment mounted under its left wing.

This was Pohang, one of the few air bases still in operation in Korea. Its runways consisted mostly of gravel, and the land surrounding them was mostly mud. Bill figured that earthquakes and tremors had caused a lot of the damage, and a few bulldozers might come in handy—but there weren't any. Surprisingly, they had been told Pohang's strip was one of the better ones. Bill couldn't believe any planes he knew about would be able to take off on a runway like this one. Flying definitely came second to repairing air strips. Repairs were the priority, but where were the parts and equipment? Possibly a few men with shovels; he had to figure it out.

Bill was then introduced to his plane, the one that should be able to take off on a Pohang airstrip. He was supposed to fly reconnaissance in a World War II era, propeller-driven P-51 Mustang retrofitted with a round, black camera secured to the underside of its left wing. He would be flying every single day, filming both sides of the 38th Parallel for as long as he was here.

He took off, and what he saw lifted his spirits more than he could ever have imagined. The Korea that lay below him was beautiful. One trip in the air took him out over trees and wilderness settlements that other people would never see, a rugged coastline and fields with temples dotted here and there. That part of him that was still a good Catholic wanted to see what a Korean priest looked like, to meet one.

From that day on, he flew his photo reconnaissance missions over ancient cultures and across borders, past great rivers and around military installations. He crossed over the 38th Parallel and back, day after day, and filmed everything he could: soldiers on patrol or armament factories. When he wasn't flying, he was Powerman again just as he had been at Eglin, but this time in Pohang. The main difference here was that there was no warehouse full of supplies; he had to pirate spare parts from downed planes, which meant lost pilots.

All through the fall, he flew his Mustang up high, then down leg to make his approach toward whatever he spotted—a bridge with troops on it or even a suspicious dark spot in the jungle. There were times he even followed soldiers walking.

All of his photographs made it back to Command. By the time his log book showed eighty-eight missions from Pohang, all of Korea was covered with heavy snows and ice, and there was little reconnaissance that could be done effectively. As for repairs, it took at least three times longer to complete repairs on Korean airfields in the best of conditions than it ever took in the warmer climes of Eglin or even in Europe during World War II. Conditions here were too bad to do much of anything.

When baby Peggy came into the world, he felt awful. Worry hit him again, as hard as it had hit him back at Eglin: he shouldn't be here; he should be home, doing more for Virginia; they probably made their decision too hastily, but what else could he have done? He would have been sent to Korea no matter what they had decided. He knew servicemen everywhere were asking themselves the same questions; none of them could ever have prepared themselves for fighting in South Korea.

Bob and Bill spent many a night over supper agonizing. Did they really have no control over their lives, no control over where they might go or what their future might hold? Would they have to be content moving forever from base to base, happy to have even a few weeks in the States? What about the people left behind?

At night, after they'd landed safely, the guys shared bottles of Green River, which was a 180-proof hooch that came from Japan.

Bill missed Virginia, but consoled himself by talking about her time at Eglin. "At least we had a little time together in a warm place," he told Bob. "I think she liked Florida. Guess I was lucky."

"You were. I wouldn't mind going back there myself," said Bob.

Beavers summed it up. "What was I thinking? Guess I swapped ice cream on the Gulf for ice in Korea."

Conversations more than likely turned to what was happening in the rest of the world, the world of communications and rockets and radar systems reaching distances unheard of before was top on the list. Eglin had taught them well.

Korea, however, was no stranger to intercontinental guided ballistic missiles. While Bill was there, an American JB-3 missile destroyed thirteen bridges in North Korea, and the troops expected they would soon have the Snark, another guided missile that could take out a rocket in mid-air without a gun. The delivery of a Snark had been delayed because guided missiles had to be guided from somewhere, and somewheres were sadly lacking in cold and hazardous Korea.

There were lots of rumors. One that took on traction was that the U.S. was building a bomber base in some barren place like the Arctic Circle, in order to have an early warning system to protect countries like Denmark and Canada as well as Alaska and New York from a Soviet ICBM. Little Denmark would welcome any such protection as would the United States, and other countries in Northern Europe. But the States could not get out of this police action in Korea just yet. America's resources and efforts had to go toward ending the conflict.

North Korea had fewer worries; they could send missiles anywhere they wanted, if they had them, because the United States had no way to track anything coming at them. Not that the United States didn't have plenty of capacity for building such machines, they did, but so did the Soviets. Everyone knew communist governments were hard at work on such machines, so the Western world had to do the same. The race for the first and most powerful weapons, both offensive and defensive, was on.

As for life in Pohang, whenever there were pauses in the fighting, each of them—Bob, Bill, and Beavers—made plans for building careers out of something important in the civilian world, reasoning they had long ago fulfilled their obligations to the military. They began their old habit of writing requests to leave the service. Bill wrote his for the third, fourth, and even the fifth time, but all his requests were turned down. Then, one day, Bill was called into an office and handed a letter from his mother, a letter that had been through many hands and bad weather. It was already open.

"She says she wants you to come home. Is she ill?" The captain had read it, but waited while Bill read it through. It didn't really say she was ill, just that she had health problems and things were not good and she needed Bill at home. Before he answered, Bill tried to consider what his mother was really asking for, but his mind wouldn't let him stay there. It ran to how few letters he'd received from Virginia and how, in the ones he had received, Virginia no longer complained about taking care of a child by herself in such a desolate place. If he were reading her letters right, she was becoming used to living without him. She'd made friends with a neighbor who helped with Peggy, and she had a job at the local telephone company.

Of course she wanted to know when he'd be home, and now his mother did, too. The word home by itself made him even more determined to leave, but he felt helpless. All he could do was recognize that he and all the others had no control over when or where the military would send them or when it might let them go. He made his plea anyway. He really did need to go home. But when the captain reminded him of the nine months of overseas duty he owed his country and suggested that enemies of Western nations could not yet sleep, he knew he was being denied.

"If I might ask one favor of you, Sir," he asked. "When you respond to my mother's letter, which I hope you will do, please mention that the USAF will not grant me leave due to overseas assignments."

The captain agreed, and Bill felt the knot in his stomach calm down. He had enough trouble as it was, and there couldn't be any war colder than this one in Pohang in November. It was as miserable as any place on earth could possibly be. He brushed off the crazy rumors of clandestine missions in Greenland and army posts stockpiling ICBMs in case of a nuclear attack from the USSR. When North Korea took the last of the decent airfields at Seoul, Bill knew America's winning streak was over. In a moment of clarity, he sat down to write his captain again.

"Who're you writing now?" Bob whispered as loud as he could from across the room.

"I'm writing another resignation letter."

"They won't accept it."

"I don't care; I'm writing it anyway."

"Hand me a piece of paper. I'm going to write my own."

That was the beginning of the end of their unsuccessful three-person campaign to go home. They would serve out the rest of their extensions according to Truman's call-up as part of what the president called the "Tactical support wing in Hold Research."

CHAPTER 5

Home Away From Home

He bought a beautiful Japanese doll for Peggy on his trip to Japan, but there was no way to mail it. Because of the bomb, Japan's infrastructure was mostly destroyed, and its people were dying from radiation poisoning. Postal facilities in Japan were a shambles, rubble everywhere.

Bill wrapped the toy parachute and the elegant Japanese doll as best he could to take back to the base so he might send them home to his family. Worst case, he would take them home with him when he had leave. Peggy would be old enough, he thought, to play with the doll. If not, he would just throw the parachute up in the air for her and let her watch it fall. The parachute was made in France out of pure parachute silk, and it even had ripcords. For a brief second he worried that such a toy might be better for a boy, but then again, he could picture Peggy clapping her hands as the parachute unfolded.

He was daydreaming, hearing Peggy's happy baby squeals and seeing Virginia preparing to move to the beach, when Bob broke the spell. Bob had walked out to the airfield to tell Bill two things: Northrop was making progress on its unmanned Snark—a nuclear capable aircraft—but it was not yet ready to be deployed in Korea. Same story with the Sidewinder, which was a missile. However, the good news was that another shipment of parts and supplies would be landing soon. They needed to be ready to work on the tanks and the heavy stuff.

Sure enough, supplies arrived early the next morning, and Bill and Bob set to work. As for Bill, he was now doing for the Air Force what he loved most: fiddling with the latest, most formidable gadgets the military could devise—satellite-linked airborne warning systems, stealth airplanes, satellite-linked aircraft control systems, precision-guided missiles and of course, the Snark and the Sidewinder. He loved U.S. inventions and trial-and-error Air Force experiments. There were plenty of training facilities for American troops in these extremes of cold, and he guessed new men were already training for Korea in cold places like the Soo, with its 30-below-zero weather, or anywhere else American troops might fight.

Bill's memories of working with Mr. Marriott's radios and air conditioners in Michigan now seemed like child's play; he was sure they were out of date by now, but such machines always seemed to improve and reappear. He saw this new shipment as an opportunity to make small improvements on control systems and could envision some powerful gadget he could hold in his hands. That last thought brought him up short. His dreams were getting ahead of him. He hadn't even finished what he started at the Engineering School at Colorado State.

The Air Force did, finally, give Bill his papers and send him home, but not Bob. Bob's Tactical Support Wing was replaced by the Eighth Fighter-Bomber Wing, and all Bill knew was that he would have to stay on in Korea a while longer. Bill and Beavers and several of their friends did everything they could to get Bob released at the same time, but it didn't happen.

Then came the letter from the master sergeant at Fort Warren, where Virginia now lived. It said "Powerman Bill Brosco of the 3454th Training Squad is hereby wanted for duty at Command Headquarters. He is considered to be highly qualified for a specified duty now assigned to him."

"Another duty? This war is still on. What the …?" asked Bill.

"They didn't tell me," said the captain. "All I know is they're taking you out of Korea. Lucky you."

On a frigid day in November 1950, Bob took one last picture of Bill in uniform standing in front of his PF-51 Mustang before they parted ways. Both knew never to say goodbye, just "see you later," the farewell that servicemen always gave one another, but he was grateful for his reprieve. Furthermore, he had served as a lieutenant and flew in Korea for many months, and he had turned twenty-one without being shot.

Bill was already standing in line waiting to board his plane to resume a career at Warren AFB when the same captain pulled him, and only him, out of that line. Yes, he was going to the United States, but not to Wyoming. There was a plane waiting for him and a few others; this plane was carrying men to Eglin Field. After all they'd been through together, Bill needed Bob more than ever to help him figure out what was going on. He was expected in Fort Warren; Virginia would be waiting for him, and he didn't even have a chance to tell her that his ticket was to Florida.

He managed to call his wife during one of his layovers, but all he could do was give her his estimated time of arrival in Florida. He said he'd get another flight home as soon as he could. That was all he could do. She didn't take this change of plans very well.

On January 29, 1951, no-longer-lieutenant Bill landed at the base in Pensacola back in his old role as staff sergeant. It was okay by him. He'd signed up for this. The next day, he went for his separation review and the new master sergeant at Eglin Air Force Base "suggested" that if he wanted to get ahead, he should take the All Air Force Exam in Engineering. He might as well. He had at least a week at Eglin before he could leave for Wyoming. He would be gone before he got the results, or so he thought. Everything changed again.

Virginia and baby Peggy flew down to Florida to be with him, and they spent every minute together, not knowing what might happen next. They devoted their spare time to looking at possible homes all around Pensacola, even though Bill still had no idea what his "special assignment" would be. They were beginning to think they might actually buy a home nearby and went together to apply for family housing. Then came a letter from Bob.

Bob had sustained an injury and was coming home, too. The whole USAF would soon be leaving Korea anyway, and Bob would call him when he got back to the States. But there was something else Bob had to tell him before he could sign off. All the equipment that he and Bill had worked on so hard together had to be burned before Bob could leave. All of it was gone.

Bill mulled over all the work they'd done. He understood why it had to be destroyed, but it made him start thinking about his life in the military. His mind raced until he settled on "I don't want to be active and I don't want to be a civilian. I don't have a college degree, not even in engineering, so where does that leave me?"

He was exhausted from the confusion of the last two weeks and swore to put the cold days of Korea behind him forever; he would concentrate on family and forget everything else. Then, two days later, Virginia and Peggy decided to go back to Wyoming. Her house was there, she said; she was tired of renting and anyway, he had to report to Wyoming before the end of the month. It wouldn't be long before Peggy would have to start pre-school. It all sounded rather peaceful and quiet, but Bill felt as though the whole marriage was fading into the background. A good night's sleep and he had to report to Headquarters APG at the Base the next afternoon.

Morning came, and after two cups of coffee, Bill decided he needed a buddy and would take a taxi out to the docks to see if Dewey were anywhere around. Dewey took his boat out into the Gulf every day, and no matter when he left, his boat regularly brought in way more fish than any of the other boats did. He sold way more, too. Just for this morning, he thought, he would wander down to the docks and he might see Dewey if he hadn't already gone out in the Gulf.

He walked the docks and watched gulls flying overhead and pelicans settling regally on the posts, hoping to grab the discards from the day's catch as the boats came in, but Dewey's boat wasn't there; it was out in the Gulf somewhere. If Virginia had been there, he would have hired someone to take them both out to fish, and they would have red snapper for supper. But that was just another dream. He went back to the base.

In 1951, the Lincoln Laboratories' radar research and development program had just opened, and their experts were already working alongside Eglin personnel to install more and better radar systems to protect the United States from Soviet planes. The 340,000-acre Eglin site in the Choctawhatchee National Forest would be used for gunnery ranges and bombing training.

Things had changed since Korea. Something new every day, and the master sergeant pushing him to do more. It seemed he heard somebody's voice call him every ten minutes to come fix one urgent problem or another, no matter what he was doing. Bill did not object, but the endless work grated on him. Finally, he could stand it no longer.

"Why don't you ever ask one of the others to do some of these things?" he asked.

That was when his master sergeant 'fessed up.

"Well, Yank, I guess you should know. Command Headquarters is looking for a man with a certain set of skills for a special job in Greenland," he said. "And you are that man. Not only have you mastered the skills but you passed the 'All Air Force Exam in the field of engineering,' and you got the highest mark."

"Whose idea was this? I thought I was going to go back to Wyoming."

"Sorry. I'm not allowed to tell you who recommended you, but this job is one we think you're the best person suited for it. It's a really good position that will let you continue at Eglin for a little while. I can assure you you've been thoroughly vetted, and so have the others. What do you say?"

"What others?" asked Bill. If Bob and some of the other guys coming home from Korea were part of this thing, it might be okay. But the man knew nothing about "the others."

"That's classified, too, but there'll be a couple dozen of you. Don't know yet."

When Bill asked him what for, the only answer he got was "for now you'll stay here and concentrate on electronics. And test on our climatic hangar. You worked with that before, right?"

Bill slumped in his chair as his tormenter went on, and in the end he figured it might still be good news! There might be a town in Greenland that Virginia would like, and maybe she could join him after all. He knew nothing about the place, but called her to say he would be joining her in Wyoming shortly, that he would only be at Eglin a few more days—even though he had no idea how long he'd be there. He asked her to check back with him tomorrow.

She was late calling. They'd worked really late at the telephone company, she said, and she was sorry for not calling sooner. After a bit she told him she just couldn't leave; she couldn't uproot baby Peggy, and besides, she didn't want to risk losing her job. He tried reassuring her that he had a new job, and she would like it, but he didn't say anything about her not wanting to come to Florida with her husband. He blamed himself for that. He'd been gone too long; she needed time; and he still had to report to the Base Commander and would as soon as he hung up.

The Commander invited Bill to sit.

"Your scores on your All Engineer Exam were above average, Boy. Furthermore, you really excelled in electronics. We've watched you work in Korea, and if anyone can stand the cold in Greenland like the cold that you had in Korea, you can. You do understand you are still on active duty?"

"Yes, Sir."

"The USAF has told me you will have this assignment, and if you complete it, you will have earned every bit of an Honorable Discharge." He took a few minutes going over some paper work.

"This assignment will not only allow you to fulfill all of your duty, but you will be doing something very valuable not just for the United States but for the world. This is a top secret mission and requires confidentiality."

"Yes, Sir … but …

You understand we have to finalize top secret clearance for you, don't you?"

The Commander didn't give him a chance to talk but continued debriefing him about Denmark, Greenland, Canada, New York, and how the Arctic Circle ran through Greenland. This was to be a USAF base, a cooperative venture whereby many countries could protect each other from the USSR's nuclear weapons, and it all had to do with where Bill was about to be assigned.

"I see you do have security clearance, so it should be easy enough to upgrade you. As soon as that's done, we will be sending you to Camp Kilmer in New Jersey. There you'll join the others whose specialties are in other necessary fields. Kilmer will train you for this assignment in Greenland. You'll finish up here this week."

Then the Commander stamped the paper and sent Bill back to the base. At least he was right about having time in Florida. It was just shorter than he'd thought it would be.

The master sergeant was still there, waiting for him outside.

"I hear you'll be leaving us. I kinda hate that … who's going to take care of the diesel engines and the radar equipment? Why are they sending you to Kilmer? Nothing up there that I've ever heard about—just a lot of Yanks."

"You ever heard of the Arctic?"

"Sure! That's where they got places with dogsleds."

"Well, all I can say is where I'm going, they don't have any dogs."

"Secret, huh?" The two men clapped hands, and the Sergeant wished him good luck.

Bill reminded the Sergeant about his resignation request and asked him to look into it again; he hadn't gotten a response to his letters, and he wasn't too happy.

The next day, when he dropped in to see the master sergeant, the man pulled out a letter dated December 5, 1952. The master sergeant had his orders: "Lieutenant General Myers will not accept the resignation of Tech Sergeant William R. Brosco. He has not completed his two years of overseas service in his current overseas assignment."

The conversation ended there. When it came time for his final briefing, the master sergeant called him back in.

"As a starter, you now have top clearance," he said, "and you need to know that this mission is part of our country's obligation to help protect Denmark and many other nations including the United States and Canada ... from the Soviet Union. We know now how far their rockets can go, and it's just about anywhere. Then there's the atom bomb—and the hydrogen one." The man leaned in. "We need more strength in critical locations where we might be able to stop them. This is the Cold War, Sergeant. This is part of NATO's Defense Program, and it's a part that will one day involve thousands of men and millions of tons of cargo in perhaps the most remote parts of the northern hemisphere. Congratulations."

Bill could feel that little pain in his stomach growing. It was just stress. His second thought was that while he was here at Eglin, he had to take every opportunity to figure out how this small world of his really worked.

"Can I go home now, Sir? I have a couple of weeks."

"You mean Wyoming or Michigan?" The Sergeant had to think what the man could possibly be talking about. He still had work to do at Eglin, but he supposed he could fix that, considering.

"Well ... I suppose Wyoming. My wife and I have some planning to do."

"Granted ... of course."

Bill bought his ticket that day. When he landed in Wyoming, he saw Virginia waiting on the tarmac as he deplaned. He saw a beautiful brunette with a happy child in her arms. He almost fell down the steps to rush into her arms, but Virginia sort of just waited for him where she stood. She didn't reach out, but then she was holding baby Peggy in front of her for him to kiss. He took the squirming baby girl and gave her a firm kiss, then he reached out his free arm to hug his wife, and she hugged him back. It was not the warmest greeting in the world, but then they had been apart so much, and he hadn't been there when Peggy was born. It would be all right, he was back in Wyoming now, and they were all on the way home together.

Virginia told him all about her job and how her neighbor had taken care of

Peggy and that the people she worked with were wonderful! She liked working.

Bill told her he didn't really want to talk about work; he had gifts for them. He was just glad to be home; but in his heart of hearts, he knew something was wrong in Wyoming. Nevertheless, the Air Force had arranged for him to stay through Christmas.

In January of '51 they called him back to Eglin AFB for one more briefing and another climatic test before he would fly to Camp Kilmer. Virginia never left Fort Warren. As he had done all those years ago when they first married, he told her that whenever he finished wherever he was going, he would have a house for her. He would never live in barracks again.

Then came the surprise. Eglin had finally granted him the family quarters he had requested so long ago. It was a cottage within walking distance to the beach and not far from the base. He called Virginia with the good news, but it made no difference. She sounded put out and wanted to stay where she was. It wouldn't be long enough, she said, to justify making all those changes.

He'd known something was wrong while he was there with her at Ft. Warren. Like so many husbands overseas, he had his suspicions. His wife might just be in love with someone else—that fellow worker at the telephone company maybe, the one she talked about. The Air Force was obviously not through with him, but he was afraid Virginia was. His heart was not in his latest assignment, but he still had a few days to enjoy the cool breezes whipping up the Gulf. The beauty of the place began to take the edge off his sadness. The one day he went fishing, he brought in his catch and roamed the docks talking to all the other fisherman, but could think of nothing but the fact that Virginia wasn't with him.

"Hey, Yank!" Dewey wasn't out in the Gulf that day! He'd always called him "Yank," and Bill remembered the first time he ever saw Dewey. He'd said "Come go fishing with us! We're a few mates short. You know everything about electronics. You told us so. But can you gut big fish?"

After that, Bill had managed to fish with Dewey's crew whenever he could, which was not that but often because he was also responsible for the radar system that never slept. Between managing diesel for a system that monitored airways, all the way from Fort Walton to Destin, and time spent testing cold weather equipment in the climatic hangars, he had little time to fish.

Finding Dewey Destin that last day in Eglin was the best medicine he could think of. Bill told him as much as he could while they were miles out in the Gulf while Bill was gutting hundreds of flapping redfish as fast as Dewey could throw them in the hold. When they tossed the last redfish in the hold, the two men hung over the railing of the boat, looking at porpoises.

"I passed an exam," Bill told Dewey, "and because of it they're sending me off to New Jersey and from there to another job I don't look forward to. I haven't finished my overseas tour of duty."

Dewey asked him if he cared to elaborate, but Bill changed the subject. The Gulf was full of rigs looking for tarpon and others dragging their lines for grouper, and they passed dozens of boats with nets full of fish.

"Just look at that. I would never have guessed there were that many fish in the Gulf."

"Not as many as you think, Yank. I see them fishing like this every day, bringing in more fish than they're supposed to—even by-catch. Mark my words, there'll come a day in my lifetime when I will see the last fishing boat with the last fisherman chasing the last fish in the Gulf."

Afterward, whenever Bill saw boats with huge catches, he remembered what Dewey said and hoped he was wrong. Life was good until Monday. The master sergeant called him back.

"You've got your assignment, Sergeant. I can tell you the AF Space Command must want you bad, and it must be secret, because they won't tell me anything. Since this Cold War began it's been crazy—so much secrecy, secrecy about almost everything. What do you have to say?"

Bill couldn't resist. "Sir, I want to apply for my deferment again."

The Captain clarified. "Sergeant, I think you're already getting some reprieve for your duty overseas. They brought you back early from Korea for a reason, and I don't know what that reason is. I just know my job is to get you to clear up all your debts and turn in all your supplies. Then I have to send you off to Camp Kilmer, New Jersey. And I am not authorized to answer any of your questions, and that's it."

"Can you tell me how long before I leave for Camp Kilmer?"

"Okay, you're going Wednesday," he said.

There was one more thing Bill had to do, no matter what. He had to try to find Bob one more time, before he wound up in yet another place from which he might never return. Beavers had no idea where Bob was, but assured Bill he could probably reach their old buddy at his old address in Hennepin. They found an old number for him, but it didn't work. Then both men began calling mutual friends, but after all the catching up they did over Beavers's phone—ignoring all the charges accumulating for long distance calls—they came up blank.

Not until they'd contacted the last person they knew to contact did Bill finally tell Beavers that he was leaving for Camp Kilmer. He had no idea where he was going after that or when he'd be back. It would be up to Beavers to run Bob down. Beavers promised, and he promised Bill he would set up a trip for all of them when Bill returned—a trip some place they could all go fishing and hunting.

"Don't worry," he said. "I'll find him. Write us, okay?" Bill nodded yes.

The bus was waiting when he got there. Bill and four others got on, all headed for Camp Kilmer. Each one had believed he was the only one, but no. Five men from five different disciplines, with Bill being among the engineers, had been briefed. All of them knew why they were going: because of that treaty between Denmark and the U.S, where they were going: to Greenland, and what it was called: the "Greenland Defense Area." Denmark's borders were especially vulnerable, but in this strange Cold War so were borders everywhere. Countries all over the globe were in the race to protect themselves from the Soviet Union and, under

Stalin, its nuclear weapons.

All Bill could think about was that his only sin had been taking that stupid All Air Force Exam, and look what was happening. He might even lose his wife and child. What could possibly be more fitting than for the USAF to send him to perhaps the most desolate wasteland in the world, Greenland, as "Tech Sergeant, Electrical Supervisor."

"Lord help!" he said out loud. He sounded just like Dewey Destin. Then "Amen" came from someone way in the back of the bus who was probably thinking about nuclear weapons.

For Bill, the question was then and would be forevermore: "How will I ever have any say in my life when I'm duty-bound to words written in stone by entities like the Continental Air Command and NATO that I will never see?" There would be an answer soon enough.

CHAPTER 6

On to Greenland

The Westwind pulled out of the harbor at Goose Bay with a steady wind blowing under a bright sky. The old cutter seemed sturdy enough, and April's sun was definitely welcome, but any sun was welcome in this sea of ice. March had been the first month they might have had enough light to work where they were going, but they left March behind when they boarded the Westwind in Newfoundland. Now that April's sun shone longer, all the men had to wrap themselves in Arctic gear to come out on deck to see it.

"Hey, Bill! I can't see you under all that fur!" said Lieutenant Peck. "They should have let us have some of this warm stuff while we were still in New Jersey!"

Bill agreed. This Arctic gear was a vast improvement over what they had in Korea. It seemed like years since he left Korea. It hadn't even been a year.

One man pointed at an iceberg in the distance and asked anyone listening, "Have you ever seen an iceberg before?"

"Nah. Ain't even seen a glacier, much less an iceberg."

"And neither of them is an ice field. We're headed for an ice field."

"How much farther before we cross the Arctic Circle?"

"We already did. It can't get much worse than this."

"Well, I think it can. Anyway I guess we can quit worrying about what's going to happen to all our personal information."

"Yeah, I've wondered that, too. All that "Top Secret" stuff they gathered. Who gets it?"

"Probably somebody in Washington."

"What if the communists get it?"

"Cut it out, man. No point in worrying about stuff you can't do anything about."

"Yup, plenty else to worry about."

In the beginning, they'd all agreed that the sub-zero weather in the cold hangars at Eglin was brutal; then, as they worked through the hard winter at Camp Kilmer, New Jersey, they complained about that. Now they were facing the real thing, where temperatures averaged minus 1.5 Fahrenheit. There had been a lot of grumbling in 1951, but once they received their Top Secret clearance, they congratulated one another and life took on more meaning, even when congratulations turned into commiserations over what might happen should a top secret mission turn public. After all, the results of all that personal scrutiny were in someone else's hands.

They had gotten past that as soon as they went to work on projects that only the USAF could have invented, which meant weeks of intense learning from the most advanced manufacturers and technological organizations in the world. Welders, electricians, boiler operators, diesel power plant operators, electrical linemen, engineers, and tradesmen became the best of the best, equipped for anything, able to solve any problem, but they knew no more about their fate than when they began.

That is, until their Captain assembled them in the auditorium that last day and closed the doors behind them. There, without a mike, he announced that their project had the go-ahead and a code name. They were the men of Operation Blue Jay, and their team now included the plumbers, welders, mathematicians, scientists, carpenters, manufacturers and designers that would build an air force base in the most remote place in the world—690 miles above the Arctic Circle about equal distances from London and Moscow—where temperatures sometimes fell to minus seventy or eighty degrees Fahrenheit.

Along with access to early warning systems, they would build a strategically located air base that would protect not just the U.S. and Europe, but many countries around the world from Soviet aggression. Bill remembered how they all burst into cheers when they heard the code name for their mission, and how that joy lasted long into the first half of the voyage. But now, with glaciers looming ahead, the mood on the Westwind had turned somber.

Academically, they knew what lay ahead: the Air Force site in Greenland was well north of one of the very few tiny Eskimo settlements. The main one was called Pituffik by the Inuit, meaning "the place where men dock their boats." Then, many years ago when the Danes began settling nearby, they renamed it Thule. Danes and Inuit had now lived side by side in the same tiny village for many years.

They also knew that every man on the deck of that Westwind would have to pitch in to cut a landing strip out of solid ice, one big enough and stable enough to handle the biggest, heaviest planes on earth as well as a dock that could accommodate massive cutters and cargo ships that would soon arrive with all the goods, men and materials needed to complete the rest of the base. That was where Bill came in.

William R. Brosco was the project coordinator for that air strip and the group's youngest engineer. He knew full well that an assignment like this could break a man. But he would not be that man. This was his challenge, albeit a somewhat more complicated one than installing a contemporary gadget in an old house. The first project, the landing strip, had to be finished before supply ships and planes could even bring in the builders and officers. All of it had to be completed before the dark days of September rolled in, and some of the team would have to stay behind during winter to maintain it. Those men had to protect whatever had been accomplished so they wouldn't have to start over again the next year, and those men would be the youngest among them.

Bill had no illusions. He had been promised both a break and an early honorable discharge, but although he had his doubts about both, he had not only accepted the challenge but sworn to himself he would finish ahead of time and finish everything right. The team would begin by completing all the landings, then worry about building enough facilities, permanent or temporary, big enough to handle thousands of men and hundreds of tons of goods and equipment. They could do it. But as they approached the last leg of the voyage, they all began to face their own frailties: none of them had ever spent four straight months in conditions like the ones they were about to experience.

The Westwind wound its way northward on the west side of Greenland, not even stopping as they passed by the Village of Thule, then Pituffik. Binoculars appeared out of nowhere, along with postcards and yellowed photographs taken by voyagers from the past, so they got at least a glimpse of the little village that past explorers had photographed. The ship was out at sea as they passed these settlements, but they passed the binoculars to look at the charming communities with their sod houses and multicolored buildings.

Both Danes and Inuit painted their buildings in primary colors, as though bright paint could ward off dark days. Passengers on the Westwind looked for dogsleds and occasionally saw dark spots that might have been dogs, but whatever they were seeing was too far away to identify for sure. Someone calculated that it would take a week for a man to go by dogsled from the future air base to Thule. Another man commented that even if they had sleds, where would they get dogs; and even if they had dogs, they wouldn't know how to train them. It was about then that these circumstances began to be personal; reality was setting in and they began to understand how far away from civilization they really were, and that there was no going back.

The cutter broke through the ice and began making its way toward a large bay, beyond which lay a vast ice sheet with more and much larger glaciers in the distance. The Westwind churned on toward the little bay, leaving a trail of melting ice in its wake that served as a navigable, temporary river of crushed ice for the supply

USS Westwind (AGB-6) escorts a vessel through the ice in Greenland, February 1952

boat that followed. As long as they were at sea, as long as they could cut salt-water ice, the rough ice never had time to refreeze. Navigation was still possible.

When the captain called out "fresh water," all the men aboard knew the cutter had now left the ocean and was entering the final leg of the journey to where they would disembark. They also knew that for the first time in many weeks, the cutter would leave not just broken ice but a trail of drinkable water in its wake. It had been weeks, but the men were ready to harvest it! They grabbed the buckets someone had stowed in a cabin and gathered ropes to tie to the buckets. Then came the scramble to be first to secure a bucket with a rope around the guard rail and lower it down below the frozen fresh water with the hope that they could pull that bucket up full of fresh water. When the first man brought in the first bucket with all that beautiful blue water, everyone wanted a taste.

Bill's first taste of hard-earned water made him remember two cows standing and shivering in the snow in back of their farm, waiting patiently for him to break through the Michigan ice and bring them fresh water to drink. But this was not Michigan; they were in a huge, strange, silent world in which the only noise they'd heard for weeks was the incessant grinding coming from under the Westwind as it cut ice and skirted glaciers. They were used to the monotony.

On one other such monotonous day, they heard a new sound coming from the ice itself, ice far behind even the cargo ship running along in the Westwind's wake. Just like tourists, everyone gathered on board and scrambled for their binoculars.

"It's the bowhead!"

"There, just behind the cargo!"

"Listen," and everyone got quiet and listened.

"They're singing!"

"Biggest whale in the world."

"And it has two blow holes."

"It can cut through thick ice like it was butter."

"What does it eat?"

"People," joked Major Master Sergeant Mullins. He wanted to be one of the guys.

"Naaa. He's kidding, it eats stuff like plankton, just like other whales."

One of the gentle animals swam slowly beside them at a distance, and once in a while it would come close enough to the boat that men could see it blow sea water, sporadically, out of its two blowholes. The whale never surfaced entirely, rather it swam with most of its curved head and only a little of its curved back above the surface. Songs from that pod of bowhead whales calling to each other was the most excitement they'd had on the whole voyage.

So far, other than their own voices, the only other sounds they'd heard came from underneath the Westwind—the cutter's own crunching, grinding noises. Its only job, other than carrying thirty-three men, was to cut through that ice—salt or fresh ice—and skirt glaciers. They were used to that. It was not until their next to last day on board that they heard something they'd hoped they would never hear. They had just passed a glacier. The sky was clear, and ahead of them was nothing but the new, warm sun of April reflecting off the ice sheet. They were just a day out from land—or so they had been told.

The noise came from something aft of the ship, and as they all gathered at the back of the boat, they saw it: the two of the glaciers they had only recently passed not too long ago had collided with each other, making horrific, sharp, cracking noises intermingled with what sounded almost like a human screaming or moaning. The cutter rocked underneath them from the waves. The men on deck scrambled as the ship turned, just to watch, and what they saw terrified them.

The Westwind had just barely avoided being trapped between two ice fields. Not a soul spoke until well after the thundering sounds faded into the distance. It sent shivers through Bill's bones.

No thunder in the Soo had ever lasted that long or vibrated like that. The chill that ran through every man aboard was a chill Bill knew he was likely to feel again. The five men from Eglin understood the sounds of a hurricane tearing through town, but hurricane winds came in hard but left fast, and even though they can leave incredible damage behind, after it's over, it's over, and in its stead come the sounds of gentle waves lapping against a sandy shore, as though nothing had happened. But the screeching of colliding ice fields seemed to last forever. It was some time before they could once again hear nothing but the familiar sound of metal on ice and the Westwind's passengers settling down. The sun would rise. They would have light, and the seas would calm enough to begin going ashore, if they ever got there.

The next day, as they passed another particularly large iceberg, the pilot—using his instruments to monitor everything going on below the surface—announced that what they were seeing now was only one-sixth of a berg. Five-sixths of that glacier lay below them, an invisible mass of ice five times bigger than the monster above the surface.

"One hell of an ice cube," said one of the men.

How the crew managed to hold their course, nobody knew. But they all had to keep their fears to themselves and put any faith they could muster in God and the strength of that old Coast Guard icebreaker. Their biggest worry now was the safety of the cargo ship behind them. By some miracle or another, the cargo ship was still there riding the liquid trail behind the cutter, and the Westwind's passengers settled down knowing how easily it could have been otherwise.

Toward the end of their last day, they passed slowly by yet another glacial ridge. Portside and some 10,000 feet high rose a silent blue glacier. The immensity and beauty of the thing overwhelmed them yet again, but this time they couldn't hang around and watch; they were busy preparing themselves for the end of their journey. The icebreaker chewed its way through unstable melting ice like a monster engorging itself on ice cream. Thirty-five men stood on deck wrapped in Arctic gear as the Westwind approached a huge flat plane of solid ice on the starboard side where they planned to drop anchor. From there, they would drive the duck boats off the cargo boat and onto whatever shore they had, which in this case meant they had to disembark on a frozen plane.

The Coast Guard Breaker, now known as the USCGC Westwind (WAG-281), chipped away at the melting floe and made noises loud enough to bring out the stars, but there were no stars in April in Greenland. Bill stood aft watching the cargo boat trailing behind, navigating the slush formed in the wake of the Westwind. At first, the ice just formed and melted, then formed again and melted again, but as they got closer to the shore, they saw the sharpest slivers of ice begin to pile up and form treacherous-looking mounds.

The Westwind was an American icebreaker that held its own secrets. For all anyone knew officially, it had been loaned to the Soviets in 1945 as part of the Lend Lease Program and had only reappeared once in the States. In early 1951, it had been sighted in Seattle, Washington, and was about to be towed to Boston Naval Shipyard for repair, after which it was to be decommissioned. On March 19 of that year, the Navy was supposed to transfer it to the United States Coast Guard whereupon it would be struck forever from all Navy lists. The Westwind was just another relic of World War II somewhere in drydock. The public, if it cared, probably thought the Westwind might yet be resurrected to break ice for Brooklyn in the winter. The truth was that Brooklyn's winter was over, and the ancient and well-used but reliable icebreaker was not in dry dock at all but somewhere in the North Atlantic transporting a team of scientists and other workers to Greenland for Operation Blue Jay.

The Westwind had seen better days and had taken another horrific beating, but it could be retrofitted to live another day. If anyone had bothered to look for that boat, they would find neither a registration in the United States nor even a manifest. A vessel in limbo is the perfect vessel for a secret mission, and huge icebergs and ice fields are excellent hiding places.

For a while there was nothing for the passengers to do but watch. They had all been through the same Arctic survival course, studied the same maps and charts, understood how roads must be laid out and how to test for a place to land no-wind flights, but nothing had prepared them for this. But the trepidation wore off, and after all their months at Camp Kilmer—all the vigorous training, heavy classes, immersions in simulated readiness exercises—they were boys again, ecstatic at the prospect of a new adventure on that unforgiving, vast, frozen land ahead just past the glacier, and they were ready.

Each one of those thirty-three men who had been chosen to be part of this top-secret operation had mastered a certain set of skills during those long days in New Jersey. Those welders, electricians, boiler operators, and the others who had spent their first three months at Camp Kilmer perfecting those skills were ready. But they wouldn't be able to test their new techniques without parts for boilers, power-switching equipment, pumps, engines, and diesel fuel, none of which they would have until the airfield was built.

Thanks to Camp Kilmer's training on sextants, astro compasses, courses in dead reckoning and celestial observations, New Jersey had gotten them this far. They could only hope that what they had learned would prepare them enough to figure out what they had to do next in this place, but they had never really reckoned on what they would find until now.

For the remaining thirty-three in the mission, one thing was quite clear: this godforsaken spot was in a perfect position to defend Denmark, Canada, and New York. This particular flat, desolate ice floe lay in the center of a straight line between the USSR and the United States, and that was where the USAF set them down. Why Denmark named it Greenland, nobody knew for sure, but they had heard the story that the Vikings named it so they could attract tourists.

As the noncommissioned officer (NCO) in charge of electrical power stations and electrical and power plant operators, Bill's responsibilities also included roads, runways, and, if they ever had any, ground transport. Any air transport depended entirely on how much of the rest was complete. Somehow, in a place inhabited only by walruses and seals, hares and Arctic foxes, they had to bring about an electrical section with electrical and power plant operators, as well as hangars, roads, runways, and traversable grounds.

Somewhere much farther to the northeast and on another, much smaller site, lay a secret radar station that the U.S. government installed in Greenland back in World War II, when Germany was its enemy. It not only detected ICBMs and bombers from parts unknown, it served as the weather station that gave President Eisenhower enough information about weather conditions to order the fighting to go forward on D-Day.

The knowledge that there was a radar station up there somewhere gave them some comfort, but they had other things to do. They had to build an Arctic air base, and they didn't even have working radios. Bill wondered again how one person could pass one test and find himself on such a desolate place. Why him? If only he could go back and refuse to take that test.

The men took infinitely brief turns on deck trying to see something—anything—without freezing to death. Among them were Peter Kerwin, builder of airfields and dams; S. J. Groves, who helped build the Jersey Turnpike; Joe Green, construction expert, and many others with other specialized skills. Like Bill, they could only speculate on what lay ahead, but they were all quite sure of one thing: they were about to be abandoned somewhere at the edge of a sea of ice so remote that their secrets were in no danger of getting out.

Their destination lay between and in front of two massive columns of solid ice, striated towers whose wind-slashed gashes appeared deeper and darker the closer they came. What else lay ahead, other than longer days of sunlight, they could not guess. The cutter skirted the first glacier as best it could and picked up speed as it cut its way through now thinning ice trails toward an ice floe whose edge was lost in the horizon. When the ship vibrated, every man on board felt at least a little fear, even though they were supposed to be used to it by now.

Glaciers in Greenland often suffered from an occasional earthquake, which probably explained what they had witnessed the day before. But this shuddering came from an icebreaker that might be on its last breath. Bill's whole body shook from what? The cold? The challenge? He no longer knew. He felt small and, well, very human.

Then the cutter broke through! With thinner ice, spirits rose; they could chip away at that ice and, if all else failed, melt it into beautiful blue drinkable water. For Bill, it was all too much like home where two cows were probably still waiting for him to break through the ice there and bring them fresh water. Ahead of him lay a vast nothing but daylight; he could not imagine how they would be able to function in deep dark after August. Unless their fortune lay in the stars. As for a calving iceberg, he thought that if stars could hear the roar, perhaps they could talk back, too. If so, this was the place for man to contemplate talking to the stars. It was an idea Bill wouldn't forget.

They were landing. Bill tried to absorb enough of that boat's warmth to ward off the intense cold creeping inside his government-issued leather anorak. If it had been made of sealskin it might be warmer. The cargo transport behind the Westwind clipped along within the wake of the cutter, ready to launch its smaller ducks, which were boats on wheels that would haul their first and only provisions and equipment on shore. These were the first of the supplies and the only supplies thus far authorized for Operation Blue Jay. The freed cargo ship followed at a cautious distance, but not so far that the floe would have time to refreeze.

"It looks like a cat about to catch a mouse," somebody said, "with the Westwind being the mouse."

"That cat is far more important than any mouse," said another. "Life where we're going ain't gonna last long without the supplies that cat holds in its belly."

All the men watching from the Westwind felt the unease that came with the thought of being dropped off some place where extreme cold could freeze the water either between them and the cutter or between the cutter and the Coast Guard's cargo ship or between the cargo ship and its ducks. The worst possible reality would be if all the men and their provisions froze just beyond reach of shore only to be found months or years later if at all. Considering that a mean annual temperature of minus 30 degrees Fahrenheit in this part of Greenland with almost no supplies would be the worst place to be lost. It had certainly happened. Only one year earlier, another cutter had been demolished doing just what they were doing. Their cargo had to get through.

The Coast Guard cargo crew's goal required launching the ducks with all cargo on board and unloading it as fast as they could so as to rid itself of both human and material charges. Then they had to turn back before their newly cut passage froze over again—or worse yet before some iceberg moved in to block their way farther out from the Bay. They were antsy and with reason: the Bay was hundreds of miles north of any civilization, and its ice was made of frozen snow compressed through millennia and almost impossible to break through. Salt water ice was much easier to cut through. Even whales could cut through sea ice. But where they were going was solid ice as far as one could see; this factor alone made disembarkation a hazard.

The two boats steered alongside each other. One team from the Westwind boarded the smaller cargo ship and helped load and release the duck boats to make their way to the ice field with all the supplies. The ducks rolled off the cargo ship and made their way as best they could—through newly broken ice and over sharp ice ridges that marked the end of salt water ice and the beginning of fresh water ice. The first men disembarked from the cutter to help bring in the ducks. Only then did the boats dare shut down their engines for even a few minutes. They were as close to what could be called a shore as they would ever be.

As the men ventured out onto solid ice, they watched their feet as they stepped gingerly across the slender ridge of ice that marked the coastline and onto a vast expanse of frozen nothingness. The men shouted to one another as they went, then fell silent, struck by the incomparable loneliness of it.

The cargo ship had unloaded the tools they brought with them for the purpose of setting up cranes to unload the ducks. As the cranes began unloading, men were right there by them, grabbing any kits they could carry to make shelters for themselves. They dared not stand still. Anything, even the smallest patch of bare skin, would freeze instantly if exposed; the intense cold could eat through any flight jacket or any army-issued pair of Arctic pants in seconds. So they were happy to work.

Muffled shouts of men in parkas rose above the noise of the assembled cranes cranking boxes up off the ducks and dropping them off on the ice. One box dislodged itself from the jaws of that crane and fell into a bit of icy water where it refroze and would remain forever.

Even if it was only to keep warm, the men worked for what seemed like a lifetime to assemble their "Atwell huts," which were nothing more than insulated tents, fitted with coal-fired stoves, designed to hold four men. Those four men meant two senior officers and two enlisted men.

Most of the cargo unloaded from the ducks were those tents and their coal-fired stoves; bags of coal and rations came later. Every time an Atwell went up, its builders went inside to warm themselves; it was like dying and going to heaven. But they couldn't rest. They dug into assignments, preparing not for another day—although it was hard to decide now when one day would end and another begin—but for how they would live.

Once they were all there on the field, Mullins became the First Master Sergeant he was supposed to be and began dividing his squadron into teams according to rank: two officers and two non-coms per tent; four men and one coal stove per Atwell. The men were by now quite familiar with the Atwell: it was made of the same cotton army duck as the tents they'd used in the War. It was not only huge, but waterproof and equipped with bunks. The crew unloaded Atwells first.

The men from the Westwind had come "ashore" in all their Arctic gear, with nothing visible but their eyes, which proved quite a challenge when Mullins ordered them to line up alphabetically.

"Get to it, Men! I'm going to assign tents according to the alphabet—like Arnold, Armstrong, Arnold, Bell, Brosco. Got it?"

"Yes, Sir," shouted thirty-three men, as Mullins began assigning Atwells.

"Tech Sergeant Brosco! Sergeant Price! Lieutenant Peck! Second Lieutenant Deck!

The four scrambled to organize themselves into a team to lift, push or pull the heavy boxes containing tent, tent's framework, and stove plus the final box that was supposed to contain everything else they would need to survive, for however long that would be. The most important box soon became the one with the C-rations and critical first aid supplies.

They left the Westwind that morning at sunrise, which back home would have been around 5:04 a.m. They would have daylight until 11:04 p.m., or what might be midnight at home. The time of day had long since ceased to be important. Daylight was for setting up and equipping their tents; they would unload their stoves before dark.

The barren ice field began to take on the shape of a miniature town made up of pearl-gray Atwells in a frozen sea of boxes, opened and unopened, and stoves waiting to be lugged into the tents. It was a mess, but Bill could see its possibilities as a work camp.

His hands had long ago frozen, even with the Arctic gloves on; it was all but impossible to hold a pencil even inside the tent. Bill was among the first ashore because he had been assigned the job of keeping track of everything that came off of or went back onto the Westwind and its cargo ship. He had been writing in his logbook before they even left the cutter, and he had taken inventory all along, even while he and his "B" team set up their tent.

"Hey, Pascouli!"

"Me?" he shouted back. The man in front of the "A" Atwell was pointing at him and beckoning him over. Bill couldn't see anything but his eyes, but he recognized his voice. It was Arnold. Arnold was standing in front of his, Arnold's, Atwell surrounded by a stove waiting to be let in and a plethora of open but not empty boxes. It was an overwhelming mess. Some of the debris could go back with the cargo ship, but that ship had to situate itself in the Westwind's wake before dark so they could leave. That window was fast closing.

"Hey, Pascouli! No bathroom supplies!" said Arnold.

"... and no cooking equipment!" added another man just coming out of another tent.

Bill would be called "Pascouli" until somebody else dreamed up a different nickname some other day. Nicknames changed with the tides or when they could be made into a good joke. Jokes never offended anyone; jokes just helped them muster the confidence they didn't really feel.

"No toilet paper!" said Deck.

"... and no cook," said Peck.

The more they unpacked the more necessaries they learned were missing. Bill, seeing they were probably all in the same predicament, asked Sergeant Price to help him double check everything. The men were right: stoves but no pots and pans, no toilet paper and nothing for their personal grooming. Sergeant Price and Bill huddled, then Bill told Mullins, and Mullins called them all together.

"You men! All of you, attention! As soon as you know where you're going to put your tent, I want two men from each tent to walk out on that ice at least fifty feet away from the outermost sleeping area and stake off a patch of ice."

The men looked puzzled.

"Don't look at me! You're going to go out there and stake off a spot for your tent buddies. That's going to be your latrine; now get to it!" He turned back to Bill, who was waiting to give him the list of missing equipment.

"You, too."

Bill, still the unfortunate non-com, was responsible not only for all engineering systems but also for all equipment and all contractors, missing or not, from that moment henceforth. He had to make the final check on everything from specifications, power stations, roads, runways, and grounds to the largest buildings, and there was no time to waste. At least one man from each tent followed Bill around, giving him their own lists of what they needed.

By the time he'd gathered all those lists of missing items, he barely had time to take them to the boat's captain before it was time for him to depart. As for the missing items, he couldn't decide whether the Air Force never intended to send them to Greenland in the first place or whether those supplies may have been in the box that fell off the cargo ship and was now frozen at the bottom of the bay. He would never know, but he didn't care. He just had to be sure those items would come back with the next shipment, which the Captain had assured him would arrive "in a few weeks." Bill took his lists to the Captain.

"Captain, begging your pardon, Sir, but we have been short-changed."

"How so?" The officer took the sheets of paper.

"We took inventory as we unloaded. Stoves were supposed to come with the ship, but we don't have enough stoves and there's no kitchen equipment at all. No bathroom supplies. No soap. Not even toilet paper. The boys are trying to wash themselves with snow. It's disgusting."

"You don't say," said the flustered officer.

"Now you tell me, how soon do you think we can get some of these things?" asked Bill.

The officer shrugged. He repeated himself. There would be a convoy, he said confidently, that would bring all of it "in a few weeks."

"Sir, can you be more specific?"

By then, the Captain had had enough. He turned to Bill.

"Your orders are only one thing: be ready when that convoy arrives!"

"We will be ready," said Bill, and the Captain turned to go.

The squadron watched until the now much lighter Westwind broke a new path out into the Bay and waited, grinding away at the ice all the time, for the cargo landing boats to pull away from shore and head back to their mother ship. Most of the men stood waterside, listening to the noise of cranes and gears grinding as the Westwind moved away. They stood there the whole time the ducks slugged their way from shore toward the cargo boat, which would follow the Westwind on the long journey home.

All Bill could do was to report back to Sergeant Major Mullins the extent of the supplies that were actually missing and that nothing was going to change much until the next convoy came. And when it did come, they must have already built not only a place for larger ships to dock but a landing strip constructed for the heaviest bombers in service anywhere.

The men would have to adjust. They had been told to be ready when the next cargo ship arrived, but the only things ready now were the men themselves.

Map of Greenland, Aage Gilberg

CHAPTER 7

Days on Ice

In the armed services one sometimes meets a lot of bad guys at the top, but Mullins was a good master sergeant, a likable tough-guy. He bragged a lot, especially about the time he directed one of the crews that built the Burma Road back in '42.

"God damn it, that Burma Road was 1,107 miles long and the most remarkable engineering feat in all of World War II, and it took us three years to do it!" He might have been right. That was before Hiroshima.

Rebuilding that road in Burma, after Japan bombed it into oblivion and shut down the Allies' only way to get fighting equipment into China, did take three years to build—in some of the nastiest, most malaria-ridden, rough terrain on earth. Mullins had bragging rights. Japan was one of the West's most lethal enemies in those days, now it was the USSR.

Mullins never gave an order without a few cuss words either; the men heard every one of them on the day they landed. Once the entire team of Operation Blue Jay was present and accounted for, Sergeant Major Mullins gave them their individual assignments and ended the meeting by calling out "Understood?" The response was a chorus of "Yes, Sirs," but when they didn't move fast enough, he ordered "Eyes front!"

That did it. The remaining thirty-three turned their attention away from trying to survive to the work at hand. Work was their only lifeline. Reality kicked in: there was no way in the world they could ever leave. It was the strangest feeling Bill had ever felt.

They had to do whatever was needed. They unloaded the coal stoves that came standard with the Atwell and set them up inside the tent. Only then did the men in Bill's tent realize that with no kitchen and no cook on base, they had to do the cooking—even though most of the food the Westwind and its cargo ship dropped off didn't need cooking. Nonetheless, even those bags of tinned spareribs, canned stew, mashed potatoes and C-rations, like those in Korea, had to be prepared. What little meat they had was all beef, no pork, and all of it was frozen solid. Bill looked at the quantities and figured that at most, that meat might last them a couple of weeks.

That first work day was pronounced over while the sun was still high in the sky. All of them were exhausted and hungry. In desperation they put their C-rations on their beds and sat on them to thaw them enough to eat. They ate and were called back outside. It had only been a break, but it revived them enough to keep going. Activities never stopped. They slept intermittently in their clothes, between orders, then went back to distributing supplies then digging holes in the ice to store those supplies outside their tent.

Inside the tents was chaos—assembling framework and untangling blankets. When word came that coal cartons were being opened, men abandoned the indoor chaos to be first in line for their tent's coal rations. Bill's Atwell Four had eaten their thawed rations thinking their day was over, then had to be first in line for their heavy box of coal, which they secured with rope and dragged back to the tent.

Only then did they become aware of what was happening around them: some of the tents were more or less floating in an icy slush that began when warm bodies went inside those tents. They dared not put heavy boxes of coal in their tent. There was only one solution: remove whatever they could from the tent to keep the main floor above water. Now they understood why they'd been instructed to dig those holes. Until they could figure it out, they would have to stow everything under the ice and thaw it out just before they used it.

There was nothing like living in close quarters with that many men for weeks on end to get to know a man. The men cordoned off by number and assigned to Bill's Atwell included Bill, Master Sergeant Price, Lieutenant Deck, and Second Lieutenant Peck. The temptation to call those last two "Hey, DeckandPeck!" was great, but they were pretty good sports about it unless they were trying to sleep.

Nicknames changed, especially when they had the potential to become a good joke. Jokes never offended anyone in that strange place, instead they helped a man muster confidence he really didn't feel.

Bill kept all the engineering drawings in a special waterproof container. In it were all the instructions he should need for building the docks and the runways, both of which had to be completed before the next convoy could arrive. This meant runway lights, markers and all, plus materials for a docking system, but the Air Force had never seen that ice field, and they came up short here, too. Bill had to decide where things must go, and precisely where would be far enough away to leave plenty of room for the rest of an air base. He had to position the runway and the airfield to be accessible to that base, and most importantly, he had to construct everything on the base far enough away from the water that even a moving glacier in some far distant future might not turn the whole base into a field of debris.

They did whatever surveying they could manage on a hard-packed ice floe with nothing permanent enough to use as coordinates. They used whatever tools they had to create points of reference for surveying, and all the men responsible for the different phases had to corroborate on a spot on that ice field that might work for all of them. Wherever that point of reference would be, it had to be at least half a day's walking time to cross.

Bill took inventory and found yet another omission in supplies. This time the markers were missing. Without them he had no obvious way either to lay out the boundaries of the airstrip or to know where to place lights. Furthermore there weren't any reflective markers that could be placed to help a pilot land. All the men had were surveyor's instruments.

Never one to say die, Bill brought the workmen into a huddle, from which they emerged after agreeing upon a point to begin. Bill crossed himself as he marked the spot with empty coal carriers. Their building days—and nights—had begun.

The first morning after a good sleep, all the men in all the tents had to go, which turned into somewhat of a circus. Whenever a poor fellow had to poop, he had to race across the ice to his allotted patch, then pee or whatever and race back as fast as he could before he froze to death. This daily activity was soon fodder for many brown-and-yellow ice jokes.

In time, that first piece of discolored real estate became unbearable, and they all agreed there had to be a change. This they accomplished, with great ceremony, by moving the poop area another ten yards away.

As for the airfield, they weren't getting anywhere. Even in May, it snowed so much they were unable to clear the proposed airfield enough to begin work fast enough to do much work before another snow fell. The men worked harder and faster and longer every day before they saw any progress. Out there, in their furry outfits, they had to identify each other by their shouts. Nicknames came in handy.

When they returned to their Atwells, the conversation often turned to hunting and fishing; just looking at those endless C-rations three times a day made them hungry for something else. Anything to break the monotony of the menu. After about a month of complaints, or rather discussions, and due to the fact that they were grateful to have any food to eat at all, on the day it was Lieutenant Peck's turn to unwrap the tin of dried sardines, Lieutenant Peck just sat there and stared at those sardines. He could not bring himself to open the tin.

"Even frozen fish would be better than this."

"Sorry, but there's no water under this ice field so no fish. No land, even. Just more ice."

They knew, or at least some scientist had told them, that Greenland's ice field had neither land nor water under it. It was made up of dense, compressed, solidly frozen snow made as eons and eons of snowfalls compressed under their own weight. That ice field was many thousands of meters thick.

"What about all those seals we saw? The ones you could see from the boat on the way?"

"Those dark spots?" asked Price. "Yeah, they were probably seals, but even if they were, those seals can only live when there are cracks in the ice on the ocean where sea water can come through. Those seals can't surface without at least a little water in those cracks."

"No cracks in this ice, so no seals to eat, huh?"

"Who eats seals?"

"The Inuit do."

"I think the Inuit eat fish."

"Forget it. The chances of us even seeing an Inuit, much less a fish, is probably nil."

The other three joined Peck in his vigil over the sardines.

"I've never seen anything anywhere near us other than an occasional fox or hare," said Price.

"Anybody else have a yearning for real meat?" asked Second Lieutenant Peck, "not just canned stuff?"

"Quit kidding with us, Pal," said Deck.

Price had grown up eating rabbit stew and said it wasn't bad. Bill agreed.

"Okay, what do you propose we do?"

"I propose a contest."

"Huh?"

"A contest to see who can catch an Arctic hare. There's nothing else out there!"

"I haven't seen a single one since I've been here … but then I've been busy."

"I saw one. I'm on it," said Deck, and the rest said to count them in, too.

After the base settled down that night, following a particularly long day, all four men ventured far out into a night that was no more than darkish light to find a place to set their snares, which they'd fashioned out of the wire that had once been wrapped around cargo boxes. Each night that week, they threw dice to see which two would venture out and check on the snares. For days their traps came up empty. Then, toward the middle of the next week, Bill and Norris got their first rabbit and whooped and hollered all the way back to the tent.

Word got out; the thought of roast rabbit sent every single man running out of his tent to chase rabbits. That first undisciplined chase turned into a dangerous free-for-all. The very thought of rabbit stew could drive a man crazy, and it was every man for himself.

Meanwhile, Tech Sergeant Bill volunteered to skin the rabbit before he realized that the rabbit presented a whole new problem. A caught rabbit had to be cooked. Cooking thawed meat presented a whole new dilemma—no pots and pans.

Was it Deck or Peck who suggested a shovel? They couldn't remember, but it was a logical solution: lay the raw meat on the flat side of a shovel and shove the whole assembly, rabbit and shovel, into the stove. It worked beautifully.

For a while, until the rest of the base began catching rabbits, too, all the shovels were theirs. That was when looking for abandoned shovels turned into yet another serious contest. More than one rabbit sent more than one man racing across the ice to be the first to claim one of the very few shovels lying about after a day's work. Finding a shovel meant shouting out something like "War Eagle" to let the unlucky ones know they'd missed the last shovel.

It wasn't long until everyone in the camp noticed that many of those shovels arrived with yellow snow still on them, but it didn't stop any of them from using that shovel. It did, however, put an end to any questions about where that shovel came from or what that shovel might have dug before it became a frying pan.

Perhaps it was no more than just to keep warm, but the men seemed happiest when they had an invigorating task at hand, and that would be building a runway on the ice, which encompassed some of the most rigorous aspects of road-building, like digging. They had to dig into solid ice and create flat, sunken areas that would not melt and refreeze in the summer, if at all. It also had to include a roadway to connect it to the center of camp. Men who were assigned to other, more menial duties, often volunteered, and work went on apace.

It was May, more than a month had passed, but the ice still did not melt. Furthermore, they continued to have snowstorms. As fast as they dug, the faster the snow fell and froze their work, which they now were afraid would never end. During those first weeks, men dug out new snow and chased rabbits. They were always hungry, always cold, and very dirty by the end of the week. Showers were out of the question, and they still had to build some kind of dock for incoming ships. They marked off the days, thinking they would soon see ships arrive with all the missing materiel.

Far out in the bay, nothing changed. Sea ice continued to break into sharp shards with every wave, and those shards piled up into treacherous mounds and moved as masses closer to the floe that was their shore. And at that shore began the acres and acres of solid, unyielding, slick ice that would become an air field.

Just going to the latrine was dangerous. When they had their first really big snow storm, it was a blinding snowstorm in May, and every man looked more like a ghost in the distance than a human being nearby.

Bill shuffled out of his Atwell one morning to see several such ghosts in the distance. The snow fell so fast that it mounded up all around the tent. He went back inside and downed his coffee while he waited for his fellows to assemble. There were four in each of the Atwells, three in the officers' tents. As each man came out, he shouted his name as loud as he could; nobody could recognize him in his gear, much less in this blizzard. Master Sergeant Mullins was trying to call roll.

The wind blew so hard across that ice field that every man on the base came out of his tent into a blinding snowstorm, and this was a day in the middle of May. When the last man was checked, Mullins only counted thirty-two. Everything went quiet. He called out as he walked among them.

"Any men left back there?"

The troops' answers were always "no, Sir." He vetted his list again. Danvers was missing. The last time his tentmates saw him, he was going out to pee.

"We'll find him," said one man. "I think I can find him; I saw which way he headed," said another.

"I saw somebody walking southwest … I think," said yet another, and in the end, two men volunteered to go look for Danvers.

Hours passed. The base waited, and Mullins called roll again as the day wore on, none of the three had come back. Days and nights weren't that different in May, so Mullins called roll every so often. The men retired, but when they woke and Mullins called roll one more time, all three were still missing, and piles of snow six feet deep had formed in places.

Several more volunteers decided to tackle the weather and search one more time, but this time they would connect each other together with ropes so they could look for Danvers and the other two without getting lost. Those men returned, and Mullins called roll, but all three were still missing. They were never found; and then there were thirty. Life had to go on.

Men had long ago stopped shaving. They could have shaved if they wanted to, and they did try, but it was just too difficult. The only way they could get water was by melting snow, and they needed that desperately for cooking and cleaning "stuff," so they did very little of any of it.

To Bill's surprise, and to everyone else's, he who had always been a blonde grew a red beard. Between beards and mukluks and heavy hoods, they all looked alike except for eyes, so they had to learn to recognize each other all over again by their eyes.

For some reason later on, the men dropped Bill's nickname, which had been Pascouli, and began to call him Charlie. It had something to do with the red beard, which was the only thing anybody could see when he was in full gear—which was always. Otherwise, except for those in the Atwell, nobody ever saw his regular blonde hair grow long. To the Atwell Four, he became "Two-Tone Bill." "Charlie" made no sense at all, but Bill accepted it.

Early in the project, there had been a day when the men were growing tired from digging into the permafrost, trying to clear ice and debris, they were frustrated knowing that every time they cleared a little bit of the strip for a runway, it would snow again and freeze, hard. The project only inched along, and they grew desperate to clear a really long stretch of land, one long enough for a viable runway.

They had to take breaks. One day he saw his men gathering at the edge of the construction site, pointing toward something on the edge of the camp. Bill looked in that direction; there were other human beings out there, small people wrapped in sealskins, squatting there at the end of the runway, watching the excitement. It was the first time they ever saw the Inuit.

There were groups of twos and threes at the boundary of the field, with their dogs still attached to their sleds and parked nearby. The Inuit were watching the construction crew's comings and goings as though it were a tennis match, eyes left then right then left again. When any of the workers made one especially deep cut through the ice, they clapped in unison.

For days the crew just watched the little band. The Inuit had some kind of a picnic every day; they had something to eat. Bill had been too busy to investigate, but when he got closer he could tell that the Inuit were eating something that looked like a big popsicle while they talked excitedly among themselves.

Bill and the lieutenants took a break and went over to talk to their Eskimo neighbors. Bill wanted them to feel comfortable; he wanted them to know they weren't doing anything wrong by being there, but most of all, he wanted to know what they were eating and, especially, where they got it.

The Inuit lowered their popsicles but kept on chewing. Bill had to look twice before he could believe what he was seeing. Those popsicles were hard frozen fish, complete with rough scales and tails and eyes wide open. Those Inuit were eating frozen fish like popsicles!

One of them held up his fish as though offering Bill a bite, but Bill waved a "no thank you" back at him. One of the Inuit came closer. Bill, who had been keeping his distance, decided to walk over to the one who seemed to be their leader.

"Bill," he said, and pointed at himself.

The man put down his fish, brightened, and said something unintelligible back. Bill touched his chest and repeated, "Bill," then pointed at the chief and suggested "Joe?" with a question mark, which set off a chorus of Eskimos saying Joe, Joe, Joe.

After lots of loud talk and many hand signals, he believed he had gotten it across that they weren't doing anything wrong and that they were welcome. When one of Inuit stood up and reached out to shake hands, Bill knew they had been among the Danes. After that, the Eskimos came on a regular basis, sometimes with wives and sometimes more dogs, but wherever action was taking place on the base, that was where he could find Joe and his Inuit friends. The entire Blue Jay team was always glad to see them.

In the days and weeks that followed, he developed a rapport with those Inuit civilians and tried to ask where they'd come from by making gestures north, south, east, and west off into the distance and asking "Thule?" But the Eskimos shook their heads every time; they refused to recognize the word Thule. Finally, Bill understood. Joe was saying Pituffik, which was the little Inuit village they saw in the distance from the Westwind. Joe and his family and friends had driven hundreds of miles from home, which was a long, long way by dogsled.

Bill had his hands full over the next four weeks, but as those days went on and they had more daylight, the base settled into a routine. In order to have electricity, they learned to refuel the diesel generator three times every twenty-four hours and to remove ashes every time they used their stove, which meant even more black ice. Their days were still long and consumed with thoughts of survival, but their thoughts now wandered to something more than their empty stomachs.

Chasing girls was strictly out. Sex was a big zero. It was better not to talk about that at all, much less keep magazines around that did—but there weren't any to keep around anyway. There was no mail, then there were no more snowstorms. Morale lifted somewhat. Work even seemed lighter, but it was because they were accomplishing what they set out to do.

As for contacting their families, they resigned themselves to waiting for those long promised supply ships to come in. The few ham operators among them banded together to jerry-rig a radio set as an experiment, and after they finished it decided to take a dozen men with them to climb the tallest glacier upon which to try it. They settled on the highest spot they thought might be safe, which was as high as any of them were willing to try. Then, with all their equipment in backpacks, half a dozen men set out to climb the closest glacier, to give it a try, the theory being that from such a height there had to be another such ham radio operator somewhere in the civilized world who would pick up their signal. If all went well they planned to talk to their families and find out what had happened to their supply ship.

They made it and set up their radio, then tried to patch a call through, but nothing happened. No ham radio responded and no supply ship was on the horizon. Bill put it this way, "the grand plan began well but ended with a whimper."

But they didn't give up; they decided to stay and try again. Then, on the first day of June, Bill heard the men with the short wave running toward the airstrip, whooping and hollering. They'd gotten a signal! They wanted him to come listen.

With the whole crew assembled, the chief operator tuned his set and there it was. Somebody was speaking on the other end. The voice on the ham radio said "Greenland, come in!" and somewhere in the conversation that ensued, that same voice said "NATO has been trying to get a signal from Greenland." The lot of them were flying high shouting "we did it!"

Then the voice went dark. The buzz around the camp turned first to wondering why NATO wanted to talk to Greenland, then to the fact that their secret base-to-be was no longer a secret. Operation Blue Jay was now an important and obviously newsworthy part of the North Atlantic Treaty Organization's (NATO's) defense program.

Communications worked long enough for Bill to call Virginia to ask about baby Peggy and to make another call to Eglin to confirm the fact that he wanted to live off base in Pensacola when he returned, but the ham radio was unreliable and he missed contacting the rest of his family. It didn't matter; even though there was nothing he could do, he no longer felt so alone.

By mid-June, Bill had been there 104 days. Many of the men were suffering from separations, but none had succumbed to gloom and doom yet. Bill never let himself dwell on it. If one man knew that another was feeling down, he would bring in a friend to make the fellow laugh or talk the fellow into a round of poker. There was no alcohol, not even beer, but plenty of coffee and cigarettes: good cigarettes—not just Lucky Strikes—but Pall Mall and Salem, choice smokes.

Ingenuity was a most valuable commodity. When a problem arose, like trying to build an airfield with no oil drums for markers, the men got through it together and figured it out. They decided to collect the little saplings that had pushed themselves up through that melting ice bog back in May. They harvested hundreds of them and stripped them all. Then they set about sharpening one end of each sapling to a sharp point. Next they drilled down as far into the ice as they could and, according to the drawings, stood those saplings upright in the holes. It wasn't too long before ice filled in those holes, and those saplings stood as straight as soldiers all along both sides of the ersatz runway, marking a landing strip at least a mile long. With a lot of searching, they then found the marker strips they thought they'd lost. They found marker strips hidden deep under some tenting. Now, with installations completed according to exacting specifications, they would be able to guide all pilots safely down onto the airstrip. If, of course, pilots ever came.

CHAPTER 8

Supply Ship

On June 19, 1951, a shout went out from the shore! Someone had sighted ships on the horizon. The base established communications, and every man left what he was doing to go out onto the just recently completed icy "boardwalk." They were all acting like fools and shouting stupid things like "ship ahoy!"

An armada of ships from Norfolk, VA, arrived with the 12,000 men, who would complete the base, and 300,000 tons of cargo. The men would pretty much live on board their ship during construction, at least until their quarters were built. Men from boats, ships, and the ducks on wheels—all loaded down with thousands more tons of cargo and provisions—began to come ashore. It would take days and days just to unload all that cargo before they could even dig the first shovelful of ice, much less build an entire airfield and all the barracks necessary for an air force base. Even after all the necessary stuff was in place, there would be several more phases to come, so many that it was hard to picture the finished product. Their brains weren't ready to conceive of all the offices, command quarters, radio rooms and receivers, radar, mess halls, infirmaries and clinics, storage areas, workshops or other things that would have to be functional before it could be called a real Thule Air Force Base.

Twelve weeks was a long time to wait for supplies, but finishing an airstrip and a serviceable dock out of ice before twelve weeks beat all expectations by at least four of those weeks. Headquarters was soon ready to be occupied, and Operation Blue Jay gave them the okay. They prepared for whatever came next as they watched the huge load of building materials unload.

After that, everyone pitched in to set up cranes and unpack the tons of cargo on board. Out of those boxes came new infantry outfits and medical supplies not unlike what might have come out of World War II—an entire flying workshop.

There would be some kind of celebration later, they knew, and their presence would be required. After all, they were inaugurating a brand new, incredible airstrip; and Bill, Norris, and Deck-and-Peck had all sworn not to act like the anxiety-ridden group they were. They had to be courteous and professional when they welcomed their superiors.

They waited until formation was dismissed, then scrambled back to their tent to prepare themselves. The men, in Bill's group at least, would wait in their warm Atwell until they were sure the landing was complete before they would venture out in the cold to greet the newcomers.

They were all still in their tents when a clean-cut, freshly shaven lieutenant came in to announce the arrival of cargo. The men just stared at the lieutenant from their bunks.

"He looks so nice!" they whispered in agreement.

The lieutenant took one look around and wrinkled up his nose, taking a good whiff.

"You men are unclean!"

They almost fell off their beds laughing.

"Damn, fellow," they said. "That's a wonderful observation."

They described their last sixteen weeks to him and told him, in colorful detail, what it had been like since they passed Thule Bay and were dropped off on the ice floe, how long they had been without water, facilities, toilets, even toilet paper, and how they had all resorted to using snow to wash themselves.

After that introduction, the young lieutenant suggested they clean up for dinner because they would be dining on the ship. Then the young lieutenant made a quick exit.

The first thing that came off that ship specifically for the men in the "Bill Atwell" was a relief package. Inside were smaller packages with a few toiletries. Bill and the other men did the best they could, cleaning up and even shaving, but even with the sparse new supplies from their new DOP kits, they still had to use snow melt to shave. It took a long time and considerable effort in a tiny tent for four men to improve their presentations enough to venture outside, but the lieutenant was waiting to take them to meet the Captain.

From there, they were directed to the waterfront where a sailor was standing by one of several ducks banked beside the water. The sailor had his instructions. He was to take them offshore where they were to board the first of the two ships, and with glad hearts they headed for their first encounters with real company after a very long time.

They had done their best. Relatively clean men boarded that duck-on-wheels for the trip out to sea. They knew the names of the first two ships in the convoy: the General Hersey and the General Butner. They were to board whichever came first.

Once he was on board, Bill confessed. "Let me tell you, were we glad to see you! We've not been able to communicate. We didn't even know you were coming until we saw your boat come into the Bay."

That night they ate their first real meal, the first one after way more than forty days and forty nights with nothing but a couple of rabbits and Army rations heated up on Atwell hut stoves. That meal was a humdinger. He had to look twice to believe it: his plate came with Swiss steak and mashed potatoes.

The next day, as he began taking inventory, Bill found all the parts and equipment needed to build at least a few of the barracks for the 10,000 temporary workmen, if not the whole 12,000 permanent personnel that would come later. Meanwhile, because there were no empty Atwells; all of the 10,000 would have to stay on board ship until they could build enough barracks to house themselves.

All the first-comers who had been living in tents were about to have real bunks. They would have barracks after the first building went up! That building, which was a prefab, would eventually be part of the command post, but first, it would be a clinic and a temporary barracks, and they were about to have their first visit from a real doctor. As they watched the building go up, the men could almost see life becoming more like the life they once knew.

Bill and his whole crew were in the process of settling in for another year of hard work when Bill saw Joe and his whole family and perhaps a few additional friends coming across the permafrost to greet him. Joe left the rest of his group and came over to Bill. He motioned for Bill to sit with him on a nearby pile of lumber, and he pulled out his pipe. What little there was of the existing base, Joe had declared his own, at least during visits. Bill joined him and took out a cigarette. Then the two began a conversation that consisted of what little Inuit Bill had picked up and what little English Joe had picked up. Joe must have been a fast learner because much of what he said Bill recognized as Danish. It wasn't long before Joe's wife joined him. When Bill began to explain to Joe that a Danish doctor named Dr. Aage Gilberg was coming, for some reason Joe patted his stomach and turned to his wife and spoke in Inuit. Then they both stood up and began to shake Bill's hand.

Dr. Gilberg's reputation preceded him. He was a Danish physician who, at least two or three times a year, would fly in to the Village of Thule to treat both Danes and Inuit. Bill realized that Joe had tried to tell him about Dr. Gilberg before, but Bill had not understood. Joe called the doctor "Nakorsaq," and Nakorsaq was Joe's friend.

It seems Dr. Gilberg had given Joe his first pipe, because Joe had proudly told Bill that story in what now passed for English. He learned Danish from Nakorsaq and he'd also learned to smoke a pipe, he said. And now Bill was teaching him English. He was trying to tell Bill something when he pointed to Bill's stomach:

"You see Nakorsaq now," he said.

It was an order Bill never expected, especially from Joe, but it made him realize that other people knew he was in pain. He'd kept quiet about it all year, but if Joe noticed, other people must have as well.

Evidently the Atwell three knew. Deck and Peck tried to make him drink Milk of Magnesia or something like it. Bill had been saved from that particular cure until the day Milk of Magnesia arrived with the ship. Now that he knew the Inuit doctor was coming, Sergeant Price would not leave him alone. Bill thought his chances of having the Danish doctor come this far any time soon were almost nil, so he hadn't complained. He had work to do.

On the day the infirmary unwrapped its first medical supplies, unbeknownst to Bill, the Atwell three—mostly Sergeant Price—managed to get word out to the good doctor by contacting the village of Thule on the radio and asking the whereabouts of their Danish doctor.

Because the Inuit were subject to tuberculosis, among other things, the doctor spent much of his time in the Inuit village. But there were times he would go farther out, to the most remote parts of Greenland, like Quiataq and Qinaq, to treat Inuit people. He could only reach those outposts by dogsled, and even with a full set of dogs and the fastest sled, it would take him "many days."

The doctor was a man much beloved by Danish Greenlanders, so when the radio operator answered Price's call, he knew all about the whereabouts of the good doctor. Dr. Gilberg was in Thule at that very moment. He had just arrived from Denmark because when he heard that one of the Inuit women, one he had known since she was a child and had treated in the past, was about to give birth, he had anticipated that she would have a difficult labor and had come straight from Denmark to tend to the young woman. Dr. Gilberg birthed babies, set broken limbs, buried the dead, and gave them all free shots and medicines.

On the day the radio operator called, that delivery was already under way. Bill could expect Dr. Gilberg or Nakorsaq as soon as that baby was born, which would be any day now.

The day the dogsleds arrived with Dr. Aage Gilberg aboard, Bill welcomed him to the base and showed him around. Dr. Gilberg spoke Danish first of course, but when Bill tried to explain, he switched to halting but quite good English. Bill was greatly relieved because he hadn't understood a single word of the Danish. The doc was there to hold the new clinic open and to perform checkups for as many men as he could while he was there, and Bill was his first patient.

The only room Bill could think of that might work as an examining room was the only building they'd actually completed—the one with a huge room and tables that served as the base's headquarters and design room where almost all paperwork took place. That was where project coordinators consulted their blueprints, because the building was protected from wind and snow. The tables had plenty of space for blueprints, so why not people?

Dr. Gilberg laid out his supplies and instruments as Bill went ahead of him moving blueprints off the tables, meticulously keeping them in order to save them from what was fast becoming a clinic. The doctor was weather-beaten and all business, taller than Bill and with a slightly graying shock of blonde. He asked Bill to open his mouth and set about examining him with tongue depressor and stethoscope. Not until he began examining Bill's stomach did he look at Bill.

"Does that hurt?"

"Yes!"

The doctor rummaged through his medical bag and pulled out several bottles.

"I want you to take blue pills one time in a day and take yellow ones for pain whenever you need one."

Bill tried to pay him, but he refused to take any money. He just said he was sorry he couldn't stick around long enough to see if the pills helped!

On August 18, 1951, the Base was on high alert: its very first plane was about to arrive, and its landing would mark the opening of the new airstrip. It was as much a symbolic landing as anything, and Bill didn't expect much else, but when the plane appeared in the sky, all personnel who had built and maintained that airstrip were outside, waiting as the plane drew nearer. They watched silently as it approached and held their breaths as it started its final descent and approach to the runway, which had only tree saplings for markers. A multiple sigh of relief was heard by all when the plane landed easily and taxied the mile of runway between rows of upright saplings.

There were no mishaps. The men on the tarmac stood at attention as the passengers disembarked. Among them was Air Force Chief of Staff General Hoyt S. Vandenberg, and Bill was among those whose job it was to greet him.

"Good evening, Sir, and welcome."

The General shook all their hands, then Bill took the General's luggage and took him to the first of the new structures, the only one livable enough for a guest. They rode in a Packard that had arrived with the ship. There, Bill turned him over to First Master Sergeant Mullins and let the two walk. He would see them later at the mess hall.

Later that evening, after they'd all finished another delicious meal on the ship, the General thanked them all and congratulated them on their accomplishments. Then, rather than sitting down, he walked over to Bill, thanked him for his service, and brought up a subject for which Bill was not prepared.

"Son, First Master Sergeant Mullins here says you're the best person to help me. I have a request. You see, I want you to help me find an Arctic fox."

It wasn't a question like "Do you know where I can find an Arctic fox? or 'Would you be kind enough to take me out on an expedition to look for an Arctic fox before I leave?" What he did say was "I want to find one."

"Yes, Sir. Happy to, Sir."

But all he could think about was how much still needed to be done before he could take his leave and how much time it would cost him. Somehow he had a feeling he didn't have all that much time. He definitely should not have eaten that steak. His stomach aches really were worse. Vandenburg's pilots also wanted to leave, so he had to put together an expedition for General Vandenberg as fast as he could so he could let those planes go home with Vandenberg on board.

He had only seen an Arctic fox once, when a white fox appeared just as he and Joe were sitting on the edge of the field, smoking. Joe had knocked his pipe clean that day and put it inside his coat before he walked over to that fox and left a scrap of fish nearby. The fox didn't run but waited as though it knew Joe would feed it.

A trip with a general like Vandenberg out onto a barren ice field, he thought, would require more than one sled, but Bill had no idea how to commandeer even one dog sled, much less enough dogs and sleds to go on an expedition. He could not refuse the General. Besides, he liked the man. Then he thought of Joe and figured he knew exactly where he might find him. So he covered himself from head to toe and walked way out past the runway to the edge of the field, calling Joe's name all the way, but apparently there were no Inuit there that night. The next morning, he ventured farther out, until he was completely out of sight of the base-in-progress, when he saw the little Inuit family coming his way. There was Joe, grinning. The two began their hand-signal communications with Bill doing his best imitation of a sled dog. Then one sled. Then six sleds. Joe got it, and within a few hours, six Inuit men arrived with sleds and dogs. Bill informed the General they would set out first thing in the morning. Meanwhile, to show his thanks, he gave Joe and his friends as much food from the mess hall as he could finagle.

When morning came, Bill sent out an entourage to find General Vandenberg and bring him back to the place where the Inuit waited. When the General showed up. Bill and Joe added a few layers of furs to the General's leather clothes, wrapped him snugly in a blanket on the sled, and the expedition set out to find the Arctic fox with enough provisions for several days plus enough for Inuit and dogs. He'd brought too much food: the Inuit did not like Western food and had brought their own, including their famous frozen fish popsicles.

Vandenburg was a delightful man, and he and Bill had plenty of time to become acquainted. It didn't take long before the General knew everything there was to know about Bill, and he even befriended Joe. Bill knew very little about General Vandenberg, but in spite of all his angst over losing that much time and having to postpone even for a day his own return home, he began to enjoy the trip. Being in such close proximity to a general came with its own challenges, of course, but they had plenty of time to address them.

They were off as the bitter cold wind caused them to cover everything but their eyes. Joe urged the dogs on, faster and faster. Sled teams had to have a head start in order to vault over any cracks that might form in the ice. They reached frozen Lake Alida, and Joe pointed out Brother John's glacier. Men and drivers fell silent in front of the glacier's brilliant white, its shadow on the ice. Then, as the sun hovered closer to the horizon, the interior ice of the huge glacier turned a brilliant, transparent blue-white as though there were a giant behind a curtain, managing the controls. Beneath the sunglow and between its most revealing crevices and fissures, that glacier told far better than words that Brother John had been there many millennia before any man and was not to be disturbed.

Joe slowed down beside a bit of solid ice and pointed east. And there they were: a regal pair of foxes just up ahead. The male fox in all his furry elegance watched every move they made; his mate lay beside him, blending beautifully into the background. Neither animal moved. Ears erect, they just looked back at the General.

Bill signaled Joe to stop the sled, then took out the camera they brought. It was then that General Vandenberg opened up the front of his great fur jacket and slowly reached inside to take out his gun. Joe was still smiling as he stopped the sleds and pointed out the animals he fed so often, but when he turned back and saw the General with his gun, the smile went away. For a minute the two men just looked at each other, then Joe tried making hand-to-mouth motions as though to ask the man if he wanted to eat the fox. Joe pantomimed eating a popsicle, but the General had no intention of eating a fox. He shook his head no and pointed at his fur coat. He wanted the fox's pelt. The General then took his one shot and felled the fox.

Bill kept his eyes on Joe, who pulled on the reins to hold his dogs back and walked across the ice to the dead animal. He leaned over to caress the dead animal before he picked it up and held it against his shoulder as he walked back to give it to the General. It was still a beautiful animal.

The message was clear. Now someone had to skin the animal and give the General a clean pelt to take home. Bill thought they would have to go back to the base to skin and clean the animal, but even as he fretted over it, one of Joe's men skinned the fox there on the ice and somehow rid it of all its innards, then wrapped the pelt in a cloth and gave it to the General. The rest of the fox went into a little bag so the little Inuit tribe could have fox for supper that night. Only then did the little party turn back and begin their final leg to the airfield.

That afternoon they would all, except Joe, attend the ceremony. The General had come to inaugurate the airfield, and he would have plenty of time there to show off his pelt. The convoy planned to depart the next day for its trip home.

When the time arrived, Bill took the General to his plane to see him off, and the General thanked him for taking him out to find the fox. Then he thanked the others for entertaining him during the festivities. When he waved goodbye to the men seeing him off, he promised he wouldn't forget any part of it. It had been a memorable time, a glorious inauguration for an extraordinary accomplishment, he said. Then Vandenburg boarded his plane, which still stood on the same spot where it landed—on Bill's newly finished 10,000-foot runway, and took off.

While Bill was out looking for a silver fox, the earlier construction guys finished their work and were all champing at the bit to go home. They would return next spring to finish the rest of the air force base. They were leaving operational construction behind and, although it could be used during the winter, it had to be manned through the long winter so it would still be operational when they returned to finish it. Those men left, and they left Bill and a skeleton crew there to stay on through the dark days.

CHAPTER 9

❖

Last Years at Thule

They were well into the new year. Bill had been on leave the whole month of January 1952, where he'd spent his time back in the States with Virginia and Peggy in Wyoming before he had to go. By late February he was at Camp Kilmer, updating himself on changes in equipment and ordering supplies to take with him to Greenland. By March, he was in Greenland out on the new docks before the sun came up in time to meet all the incoming planes and ships. The convoy brought with it the last big load for the air base, including the biggest and most advanced equipment the military could find. Bill had only seen that list on paper, but when the unloading began, it overwhelmed him.

Unlike the first year of rare shipments and visits from curious dignitaries, this second year was all business. They were expected to finish all of that base during the days of daylight in 1952, which meant installing and have working the most up-to-date everything the U.S. Air Force had been able to dream up, all in the unpredictable extremis that was Greenland.

Things had improved. Everything was expected to go smoother than it had when 10,000 workmen slept on board their own ships while they built their own barracks. Bill surveyed the base. It looked as though it was still in tact. The "third armada" had arrived and was unloading the trucks and other materials they'd ordered for the final buildings that would include a permanent headquarters as well as several huge covered storage and work areas, an infirmary, and an officers' club.

The men unloaded the Arctic Trucks in the middle of the night while Bill slept, but they woke everyone when they began testing them the next morning. From then on, Bill and the rest of the camp would wake up to the sound of Arctic Trucks removing waste or running snowblowers, which they did for the rest of the day every day. Those specialized behemoths had been made in Oshkosh, Wisconsin, and required at least two men per truck: one to drive and another to operate the snowblower or whatever machinery it carried. They came with plenty of space in back for cargo and men.

Bill sat upright in bed. How many thousands of pounds would those trucks carry? How much human and how much dead weight from engines and tanks and other machinery would pound away at that ice? And what could happen because of it? Even though he'd calculated psi's again and again, he'd learned his lesson: there were far too many unknowns about ice. They had all been briefed that no earth existed under that ice. It was all snow, millennia of snow-upon-hard-packed snow, and the world was just beginning to find out what that meant. Traffic had to withstand ever-increasing ice some days and retreating ice on others, and the ice had to withstand the trucks. For now, just removing snow from the mile-long, two-mile-wide air field or bringing fresh water to the barracks or removing sewage from toilets seemed to be going smoothly, but the day would come when those trucks might have to transport ever-bigger bombers and tanks somewhere beyond that icy road.

Arctic Trucks came equipped with giant chutes for blowing snow off the runways, were rated water-resistant, met military standards for thermal and shock resistance, and they had been retrofitted with axle spinners with sharp edges to handle very steep or very hard-packed snowbanks. Every day all day ice crews yelled above the racket of tire chains rattling and spinners scraping across sheer cliffs of ice. It was unbearably loud and lasted all day, but at night, thankfully, all those trucks—Arctic, fire, and otherwise—had to be tucked inside another building to keep them from freezing.

Among the buildings just completed was the Officers Club. As thanks to all the construction workers who were about to leave and the personnel staying behind, the officers planned a party for everyone, including the non-coms, at their new club. They were going all-out, catering it from the mess, serving booze, and playing some kind of music they referred to as "diesel smoke."

Bill was in charge of the keys to the beer on the Base, but when Sergeant Major Mullins came knocking on his door, Bill had no time to go with him to get beer, so he decided to give Mullins the key and have him go over to the warehouse and bring some of that beer back to the Officers Club. Because Bill was responsible, he wanted to make sure Mullins would be accountable before he handed him the key.

"Mullins, when you go to the warehouse, I want you to count all that beer. Not just the beer you take out, although I want a count on that, too, but I want you to leave at least a hundred cases there in the warehouse. There's plenty for the party, but I am going to count them."

"I'm sorry you don't trust me," he said.

There was a lot of beer that night. Probably more than enough to make the party the rip-snorting success it was, but Bill still had to account for everything.

"Thanks, Mullins. It was a great party, but I know you are a rascal and I am going down there and count those hundred boxes myself." They were sort of half ribbing each other.

When Bill made his count, it was just as he had suspected. He counted and counted again. Mullins had only left eighty in the warehouse, and on top of that, he was sure there had been some left over from the party. He would deal with that tomorrow.

Only a few days later, the Officers Club caught fire. The new fire trucks got there in time to put it out, but the damage was substantial. The main part of the clubhouse burned to the ground; the only thing left was the warehouse with the eighty crates of beer still in it, but the damage was so great it took Sergeant Price and the others in the former Bill Atwell many hours to salvage every one of those eighty boxes of liquor and beer. With a little help from Mullins, the five men loaded it up in a truck and buried it for safe-keeping in a snow bank just beyond the edge of the field. With no count on file, that liquor was technically contraband.

As personnel left for their homes later that year, the base began to have a deserted look about it. Bill was just beginning to relax and think that everything was going along smoothly when Master Sergeant Mullins knocked on Bill's door. It was close to midnight.

"Hey, Pascouli! You wanna go out and plow some snow tonight?" Mullins was already dressed in his furry gear.

"Sure. I don't have anything else to do," said Bill.

They walked out to the airfield where Mullins had one of the Arctic Trucks waiting. He took the driver's seat, and Bill sat where he could manage the roto wing and the chute that blew snow off the runway. They took off laughing, letting the snow fall where it would, including inside the truck the whole time they flew across the runway toward the snow bank on the other side of the airfield.

There they were, two good miles away from the barracks. So they dug three bottles of Scotch out of the banks—the kind with the little black and white Scottie dogs on the label—and it wasn't long before they were drinking that Scotch right out of the bottle, revving the engine on that Arctic Truck and whooping and hollering all the way across the airfield and back until Mullins pointed out a dark shadow walking toward them from the control tower.

He stopped the truck sharp, and a bottle of Scotch flew off that truck and into another snow bank. The man standing in the middle of the runway wore a heavy mukluk and was making his way toward them on foot. They waited, and as soon as the man got close enough to recognize, they had no doubts at all about his identity. It was the MP; they'd forgotten the base now had MPs.

"You men need to answer some questions," he announced as he got close enough to shout. The two men capped their bottles and hid them under their jackets.

"Do I know who you are?" asked Mullins. "All I can see is your eyes, and not much of them in this dark."

"Men, I think you've been going a little fast here."

"Oh no, Sir, not us."

"Yeah. The control tower saw you guys marauding all over the airfield. What are you trying to do? Tear it up?"

"On no, Sir. We got a whole lot of snow off that airfield, and it's in perfect shape now."

"I got a report from the control tower. You guys had to be going more than 60 miles an hour!"

"No way. Thirty-five tops."

"No matter, you need to file a report."

"Yes, Sir," and they did it with their capped bottles tucked inside their coats.

Bill and Mullins checked in with each other a few times those last few weeks, but nothing was ever said again to either of them. It was probably too close to going home time and that MP must have decided they deserved a little celebration. They were also too close to finishing the last of the barracks, and they needed every man.

The panels left over from last winter had to be assembled. They were made of the same kind of building materials Mr. Marriott used in the walk-in refrigeration units back home in the Soo—the same old Clements panels. Quite naturally, the designers had to give them a new name. At least while they were used in Greenland, they would be known as Arctic Panels.

Construction never seemed to stop. Melting ice or no melting ice, buildings went up. Bill and Norris stood at the construction site watching the last of the work unfold. Men with business elsewhere on the base were already coming in and out of their new barracks, and those who hadn't yet left their Atwells were retrieving gear and hauling it off to their new digs. Most had settled in their new "barracks on ice" by now and had stopped patting each other on the back as they'd been doing for days in celebration of not having to live in Atwells.

Bill finished what he had to do and, because he was feeling that old pain in his stomach again, was only half-watching the comings and goings when he saw a problem. His brand new Arctic Panels were already in trouble.

"My God, Norris, look at that!"

Norris was sitting on a tool box and turned around to look. It only took a second for him to see the same thing Bill saw.

"It's the heat! Those men have turned on the heat!"

The Arctic Panels, which had been treated with waterproofing, were growing dark shadows from the ground up. The barracks were soaking up water.

"Right. And the permafrost under them is melting," said Bill. Bill couldn't repeat Norris's answer. It wasn't in his vocabulary.

"And what do you suggest we do, Bill?"

Bill suggested they consult with the crew's overseer. Together they could solve any problem. Master Sergeant Norris volunteered to go find the overseer but suggested they bring his crew in, too, so they'd have more than three heads. He found the crew's new supervisor absorbed in unloading a truck.

Norris tapped him on the shoulder. "Hey fella, looks like we've got a new problem," and he took the overseer over to the building site. "Any ideas?"

"Not me," said the overseer. "How about the whole crew?"

"I'd hoped you'd say that."

The overseer put his fingers between his teeth and whistled, and all the men heard him and turned to look.

"Stop what you're doing," said the man. "We'll have to take a break until we can figure out how to fix it."

"We gotta buy all new panels," suggested one.

"That's not gonna do nothing," said another.

"Paint 'em with more waterproof" got a "It'd still warm up, man. It'd still melt ice."

"Not in the budget," said the supervisor. Then he turned to Bill.

"Your problem, Sir. What do you want us to do?"

The man was right. All problems having to do with building anything were, in the long run, Bill's problems.

"In Florida they build up on stilts so that rising water won't destroy the building. Maybe we can hike 'em all up on posts, high enough so cold air can flow underneath and not melt the permafrost."

"Begging your pardon, sir, but we don't have any posts."

"Okay. Stop putting up more buildings until we get some." With just a tinge of pain in his stomach, Bill knew he had to stop work. Stopping work for any reason was against his principles. As soon as he'd figured out the necessary specs, which turned out to be posts at least as long as the building was high—which was two floors plus—and heavier drills to drill half those posts deep into the permafrost where snow would fill up the holes, once the posts were in place and the snow turned to ice. Posts dug deep enough into the permafrost would solve everything.

The radio operator put Bill through, and he placed his order. The clerk at the warehouse in the States promised they could have everything on a ship and they would receive all their materials in two weeks. He expected him to say that. He'd heard two weeks before. Nevertheless, he stopped work; the crew would just have to wait for their ship to come in. It was back to the Atwells for some.

By the grace of the construction gods, pilings and drills actually arrived in two weeks. They drilled those pilings deep, and when the barracks began to dry out, the rest of the Arctic Panel barracks went up almost overnight.

In a way, they were not unlike Florida stilt houses. Those houses were built on stilts to protect them from the winds and waters of hurricanes, and they had withstood many a hurricane even while other, finer buildings were destroyed. Bill couldn't help smiling. Those Florida houses-on-stilts came in one of three configurations: A-frames on stilts, Dutch-roof homes on stilts, and cottage-style two-bedroom homes on stilts—all ordered as kits from a Sears & Roebuck catalog for about $900.00.

Of course, Arctic panel buildings had to be far bigger than any Sears & Roebuck house; they had to house ten thousand men. Still, whenever one was finished, the construction crew stood back, admired their handiwork, and again slapped each other's backs. They had a right to celebrate. The camp had come to life, but 1952 would end, and some skeleton crew had to remain. By September, many of Bill's buddies and most of the construction crew had left. By mid-September only 400 men remained, but they would be there for the rest of the winter, keeping snow off the runway and ice at bay

Dark came. Only the thinnest twilight separated night from day. By October, the winter skeleton crew had more or less settled down to the dark, but there were times when Bill felt as though his whole body was fading away, lost even, in the cold and dark. He kept telling himself that he was not the only one; all the men who had been left behind with him must be suffering. The loneliness of total darkness had to be overwhelming for all of them. They each confessed, one way or another, how hard it was, but they kept on going.

It wasn't until one day when Bill was making trips back and forth between his work site and the supply cabin, that he saw a man in nothing but an overcoat bending over the rear wheel of a firetruck. He thought to himself that the man must be cold and there couldn't possibly be anything to see on the wheel bed of a firetruck, but he put it out of his mind and kept going. A little later, as he was hauling bits and pieces back to the worksite on his return trip, he looked for the man. It was quite dark, and Bill didn't see him at first, until something made him look down at the ground. There he saw the man's boots sticking out from under the truck, but Bill could not see the rest of him. He had to bend down and shine his flashlight under the truck in the man's direction.

"Hey fella, do you want to come out and talk to me?" There was no answer.

Bill asked again, this time saying "Hey fella, come on out, I want to talk to you."

Bill stuck his head under the truck and had to keep on talking before the man even moved. He was alive, at least.

Bill reached out for him, but the man began muttering something like "they're in the ice, they're in the ice," and rolled his body farther away. The poor man was hiding from some terror and all the begging in the world didn't help. Remembering how alone he'd felt that night those many years ago when he'd had to walk the tracks to the train station alone, Bill had an idea.

"Come on out, Fella. Looks like we need some help with that ice. I don't like it either, and I promise you ... I promise you I know where we can get help."

By now the man was suffering from severe frostbite and was too weak to resist. Bill held firmly onto his arm, not sure what to do next, until he spotted two men coming from headquarters and called them over. Together they managed to pull the man out from under the truck. The poor man was unable to speak, and his eyes were glazed over. The darkness and the isolation had gotten to him.

They took him into the master sergeant's office where it was warm and where the first master sergeant would be. Only then, and after a little time to warm up and a whole lot of time cajoling, could the two convince the man that he needed help. One thing Bill knew for sure—he could not have fought that man's demons by himself. That help had been mighty welcome.

The first master sergeant radioed for an evacuation plane, and several of the men volunteered to go clear the snow off the runway. Thankfully, they'd been maintaining it all winter. The plane arrived two days later and sat on the airstrip with the engines running, waiting for them to bring the man out onto the airfield.

For two days, the terrified man had tried to fight them off, tried to escape, and by the time the plane arrived, the poor man was in restraints. The same three men brought him out to the airfield, with the poor man shouting angrily at them until he finally lapsed into confusion. They helped him up the ramp and were about to turn him over to the plane's crew, but before they did, they shook his hands and praised him for his service. When they told him goodbye and wished him good luck, the poor man leaned toward them, nodded almost imperceptibly and climbed into his seat, where his restraints were secured.

The pilot had instructions to take the man to the closest hospital they could find, which happened to be a hospital in Newfoundland. That, too, was waiting for him.

As days and weeks passed, several of those who knew him took turns checking on the man, but heard nothing other than he was still in the hospital. The last news Bill heard, a few days before he left for the States, was that the poor guy had been moved to a Veterans Hospital near Goose Bay. Of all places, Goose Bay.

Everyone knew that uranium had contaminated the St. Lawrence River at Goose Bay. It happened when a USAF B-50 bomber had engine trouble as it approached the Bay; the pilot asked for instructions and was told to eject the bomb he was carrying. Bill was just leaving Korea when it happened. At the time, neither Canada nor Labrador wanted a live bomb in the St. Lawrence River, so they agreed to detonate it, which was really all they could do. Now the St. Lawrence was contaminated by uranium-238. The authorities still had no idea what the extent of the pollution might be.

The cold and dark in Greenland had reached its apogee. Everything that could freeze at the base had frozen, which meant the construction thus far would be more or less free from environmental destruction. It also meant that Bill and the other men still there could head home for a few months before they would have to return.

A plane arrived for them, now. As they flew over Greenland, the sight was unimaginably beautiful, but as they flew over Goose Bay Air Force Base, the only thing Bill could think about was the poor guy under the truck who was now at the Veterans Hospital there, with all that uranium in the water.

The flight was an overnighter, and he arrived in Florida before he left Greenland! Eglin was a welcome sight. Virginia and Peggy had come to Eglin during his last leave, and now they were there to pick him up in Pensacola. They drove together to their new quarters: a little house painted white with dark green shutters, one of many similar ones just off Gregory Street. For once, the Sergeant had granted him his request and given him an okay for off-base housing. His family had been in the house for a few months now.

Christmas was just around the corner, and before they went out to look for a tree, Virginia had a surprise for him. Peggy was walking! Bill sat still in his chair, and when his baby made it all the way across the room and fell into his lap, he picked her up and twirled her around just to hear her laugh. Eglin had never been better; Virginia and Peggy were there with him, and it would be a good Christmas. It had all worked out just as he'd hoped it would. He even managed a visit to the doc at Eglin Air Force Base about his stomach, and the doc had given him some new pill for his stomach that seemed to be working.

Virginia fell in love with Pensacola just as he'd thought she would, and that made it a real Christmas, complete with a tree laden with ornaments and gifts. Mr. and Mrs. Beavers outdid themselves with a "welcome home" that included all their married friends from the base, many of whom now had children. Virginia seemed to enjoy it all, especially when Peggy, who stood by herself most of the night, began playing with the other children. When it came time for bed and most of the others began to leave, Bill and Beavers took care of Peggy in the den while Virginia and Mrs. Beavers cleaned up the kitchen.

In years past, he hadn't wanted to be in Florida at Christmas, but now nothing could have been better than a warm Florida Christmas with his own family. Beavers lit his pipe, and the conversation of course returned to Bob's whereabouts..

"I've covered all bases, Bill. We just can't find him. We know he went back to Korea after his injuries healed, but you know, or maybe you don't, they brought almost everybody back, and I don't believe he was among them."

Even though all the people he loved were around him now, he felt for his missing friend.

"Do you think we should go ahead and plan our fishing trip anyway?" asked

Beavers. "We can call it 'Bob's journey,' you know. Maybe that will help."

"Or make it worse. Anyway, I'll be done with Greenland in June, and I've been promised they'll give me my discharge. How about let me talk to Virginia and you to Mrs. Beavers. I know a spot in Canada just across from the Soo. We can get a plane to drop us off there, that is if I don't buy one in the meanwhile." They shook on it; they had a plan. Bob would approve.

On a very cold day in January, fish were biting. Bill and Beavers made one last excursion to the Gulf, and agreed to split their catch of red snapper and Spanish mackerel in two, enough for both families. Bill invited the Beaverses over for fried fish and hush puppies that he had made himself. He told Virginia that the dinner was part of his Christmas gift to her, and he didn't even want her to help clean up. They had far more than either family could eat, and Beavers would save the rest in his freezer to eat another day. They figured their next gathering would be to celebrate Bill's final return. Neither man said anything about the fishing trip in order not to jinx it.

The day came when Bill had to leave, and there were tears at the Brosco house all around. He had kissed Peggy goodbye amid baby tears and kissed Virginia goodbye with real kisses and real sorrow. It was hard for them all, but Bill said he had high hopes for when he returned. He'd learned, though, not to make promises other than a promise to write, which he would do even without the promise.

Then he boarded Eastern Airlines in Pensacola, which took him back to Camp Kilmer to be brought up to date on whatever the Department of Defense determined they needed to be brought up to date on. The stay in New Jersey was short. The next plane flew him to Greenland where he would finally finish the five years and nine months he owed Uncle Sam.

CHAPTER 10

❖

Cold, Cold War

The other passengers on board were civilians mostly, and some were scientists. To those headed back to Greenland, everything about this trip felt different. It had all gone too fast too soon. After his recent visit in Florida, it seemed to Bill that Greenland had only been a dream, that he'd been in Florida all his life and in Greenland only a few days, when the reality was more the other way around. By the time he received his discharge, he would have been in Greenland not quite three years. It was all going to be okay; he really did have an end in sight this time.

As soon as they landed on the icy runway, Bill headed for his quarters long enough to drop his suitcase, then made his way down to see Sergeant Norris and lieutenants Deck-and-Peck. The four spent their first day reminiscing and talking about Christmas and their thoughts on what lay ahead, which took some of the joy out of the memories of their leaves. They passed on what little gossip they'd picked up while they were in the States, one of which was about the rockets they were designing at Redstone Arsenal in Alabama.

Just because Bill had once been to Huntsville, they expected him to know all about Alabama. He did his best to explain that he worked with the radar systems and cold hangars in Florida, and they had a rocket in Eglin named Tarzan, but although he was in charge of diesel engines that ran the radar stations from the Louisiana border to Destin, he had never been introduced to the Tarzan and he knew absolutely nothing about rockets in Alabama.

Deck-and-Peck turned the conversation to how Alabama got rockets, how the German scientists brought their missiles with them when they surrendered to the allied troops in 1947. Both scientists and missiles had first landed at Fort Bliss, Texas, then they were moved to White Sands, New Mexico, where rumor had it they picked up where they left off in Germany and continued by making missiles for the United States. Nobody knew the details, but they all had questions. The big question was what were Nazis doing in the United States making American rockets, even if they were scientists?

Bill had nothing to do with that and tuned out the conversation. As soon as he settled back down at the base, he would have to pack up all the equipment they no longer needed for assembling Clements panels then take stock of what was needed for the final build-out. They had depleted their inventory for sure, and it was time to resupply the base. When he arrived, Master Sergeant Norris, Bill's best friend, said he would gladly go with him, but they would have to take First Master Sergeant Mullins along, too. They made plans for the gambit as though it would be some kind of holiday.

The nearest company that had replacement parts was in Newfoundland, and the only aircraft they could hire was an old C-34. They made their plans. Bill agreed to be responsible for all the equipment they had to bring back, but he already knew the list would include some heavy equipment that they could not take on the plane. It was up to Bill to be sure that equipment was necessary and that it would do what they wanted it to do. Preparations took the whole first week, but they were ready by the time the old C-34 landed. The thing seemed sturdy enough, but it would have been built in the mid-1930s, and Bill could only hope its owner took good care of it. All three men had their doubts, and Bill and Norris refused to talk about it. Mullins, however, made a point of it.

"Hey, Pascouli! Do you want a parachute? I can get you one for a dollar!" Mullins wouldn't hush, and Bill declined.

"No thanks! I'm not going down on any ice cap! I'll just go wherever the plane goes."

The flight to Newfoundland was thankfully uneventful. When Bill and Norris were about to leave for the Company's warehouse, Master Sergeant Mullins decided he had somewhere else to go, and after he promised the other two men he would be back in time to help them load the C-34, Mullins left for parts unknown. Bill and Norris set about placing orders for re-supplying the base. After they arranged to ship most of it on a cargo ship, they headed for the airport where they had to wait a long time for Mullins.

Finally, they saw him climb out of an airport bus, but he was not by himself. He began unloading box after box after box from that airport bus. Mullins had his own boxes, and there were a lot of them. Bill began guessing at the total weight of Mullns's cargo in his head and wasn't at all sure the old Cessna could carry it.

"How much weight do you think you've got there, Mullins?"

"Not enough to cause a problem. Don't worry about it."

Bill had no choice but to take his word. Mullins once again offered him a parachute. All Bill had to say to Mullins was that he sure was just glad the heavy equipment for the base was coming to Greenland by ship. They took off.

Sometime between take-off and landing, Mullins volunteered to unload the plane all by himself! Bill had never seen the man this generous, but when they landed, Mullins did indeed unload the plane. As astonished as they both were, Norris and Bill left him to it, but not without some anxiety. By the time the sun lowered itself to just above the horizon, which was as close as it would come to night, Mullins called them over. He'd finished unloading, and if they'd like to join him, some of the guys at the base had planned a party in the blueprint room. Norris and Bill were, of course, invited.

Bill declined. He needed to finish his job, which would take him most of the evening. Norris had paperwork, so the two men retired to their separate quarters to work.

Later, Bill hollered down at Norris, wanting to know if he wanted a cigarette break. Norris joined him outside, and they stood around smoking. They were enjoying the quiet when a wave of raucous laughter broke the silence. As Bill and Norris stamped out their cigarettes, they saw barracks windows begin to fly open, and boom-boom music and raunchy lyrics poured out those windows. All the noise came from the second floor of the newly occupied barracks at the other end of the station. By the time the decibels reached unbelievably stratospheric heights, Norris and Bill set off to investigate.

They climbed to the second floor of the barracks, which meant, of course, that by the time they got to the top step things had quieted down. They looked at each other before they opened the door as though to ask "do they even know we're here?" when someone opened the door.

Norris and Bill could see a dozen men standing around with whole unopened bottles of liquor in their hands. And there, in the back of the room, stood Master Sergeant Mullins holding an open cigar box with a pile of dollar bills in it. Mullins hadn't just bought the liquor, he was selling it! And that was bootlegging, and bootlegging was illegal.

Norris and Bill had no choice but to report it to the base. But to every man's dismay, the MPs took Mullins in and held him until, months later, there was a trial. Mullins was court-martialed and sent home.

It had been an especially sad day for Bill. Not only had Mullins been one of his best friends, but those judges refused to take into account what Mullins and all those men had been through. It went down hard for the whole base that the Air Force court-martialed their own man.

It was thus that Bill, who began as Staff Sergeant and had gotten as far as Tech Sergeant electrician, was about to spend the rest of his days in Greenland as First Master Sergeant. He was next in line after Mullins, and although he tried to refuse it, they wouldn't let him. Thankfully, none of the men blamed each other and nothing changed for Bill. He was still Pascouli, but he was also the man—one more cog in a very large wheel—who had to finish that base.

The design for that final 1,000-foot pier was still in his hands. Instructions for the pier included a number of old tanks that had been salvaged from the Gulf Coast and were being sent to Greenland to be used as an underpinning for the dock. They were to be sunk deep down into the ice floor at the edge of the bay to form the underpinnings for the dock. Ordering other equipment and concrete for that work was what had sent Mullins to Newfoundland in the first place, but Mullins had nothing to do with the tanks. In the end, it was Bill and the USAF who had to turn those tanks into a dock for the Navy to use for its landings.

The next convoy of barges was due any minute with hundreds of surplus tanks, the kind used in or perhaps left over from World War II. The workers at the base were already toughened up from drilling deep holes in the ice and installing pilings in those holes to prop up the barracks, but this time they had to be far more cautious; they had to stabilize those tanks at precise angles to hold the pier, and they had to place the steel parts just so to fit over those angles.

"Splay them piers! You gotta make 'em stable. Splay 'em out!"

"Huh?"

"This ground ain't moving. How are we going to splay anything?"

"Concrete, man. Mix up lots of concrete." And they worked with what they were given.

For the first time, conflicts flared. Orders were contradicted, and Bill had to be right there with them every minute, for every hole dug, every concrete mix and every pour, every placement—not to mention his new job, which was putting the finishing touches on all peripheral electrical installations.

About then, sirens went off all over the base. Work stopped, and soon the sirens were muted enough that someone could make himself heard over the loud-speaker.

"Air Field alert! Air Field alert! Radar has picked up a submarine ten miles off shore. Radar has picked up a submarine ten miles off shore! Men, return to your barracks."

Bill did not go to the barracks. Instead, he headed for the radio room and watched the screen over their shoulders. The six planes that had brought some of their equipment still sat on the field but were about to be scrambled to try to identify the submarine. Those planes had kept their engines running ... not because they were afraid of something that might happen in the Cold War, but because they had to keep those engines running as long as they were on the airfield so they wouldn't freeze and die.

Bill waited until the planes were in the sky before he made it back to the pier-in-waiting. The tanks, or rather some of the tanks, just sat there on the ersatz shore; only half of them were aimed in the right direction.

The whole air base waited. Radar reported that the planes were near the site, but they had to wait. Then, after about half an hour, the sirens stopped and the voice came back over the loudspeaker.

"Whale sited ten miles off shore. Whale sited ten miles off shore. Workmen. return to your post. Workmen, return to your post." Of course! Whales could break through ice just like submarines, but they were the only animal who could. The men who'd gathered at the construction site waited for Bill to say something.

"Men, that could just as easily have been Soviet submarines out there. This is the Cold War, and what you are doing may probably be the most important thing we can do to protect not only Denmark and Canada from the USSR, but also London and New York. Keep on doing what you're doing. We're going to finish!" Cheers went up from among the tanks, and cheers went out again when they saw cargo planes return to the airfield. Reconnaissance saved the day, and the pier would be finished by the end of next week.

That pier lived up to its expectations. It would support anything from cargo ships and ducks to the heaviest weapons the DoD could produce—like intercontinental ballistic missiles (ICBMs). The end had always seemed just out of reach, but this was the end. Operation Blue Jay had done its job. The base could legitimately call itself a United States Air Force Base because it had everything the USAF might need for years to come, including not only mess halls and barracks, but official Air Force headquarters. It also had its own infirmary now, complete with its own doctor. The Eskimo doctor had stopped visiting; he was spending more time with the Inuit he loved and the Danes who lived with them in the little village of Thule.

As for Bill, he signed up for the very first appointment to see the new doctor. His stomach had never stopped hurting, and he counted on this new doc to fix him up. He was just waiting for the infirmary to open. The new doctor took his time settling into his new clinic, or so it seemed to Bill. When the infirmary finally opened its doors, Bill felt better already.

The doctor's assistant was busy arranging the last load of basics, like bandages and antibiotics and such, in the new cabinets when the new doctor arrived. He took off his huge fur coat, and Bill followed him into the exam room. Bill had no sooner told him his complaint then the doc put him up on a table, the same table Nicorsaq had used, and made him lie flat. He, too, began punching and squeezing him hither and yon, and Bill tried to speak between grimaces.

"Doc [uh], I'd [uh] like to take you [uh] to lunch."

"Not until we finish."

He was all business. Doc Tremble, as he was called, gave Bill one last punch and Bill's whole body suddenly and uncontrollably contracted. Dr. Tremble looked both surprised and alarmed.

"Hang in there, Sergeant. I'm going to get you some pain medicine."

The doctor began looking back through his cabinets trying to remember where he'd told his assistant to put the pain medicine. Soon enough, he brought back a small bottle of white pills that he handed to Bill with a glass of water. They were the same kind of pills they'd given him at Eglin.

"Thank you, Sir. Those are the same pills I've been taking. Mostly the pain medicine I have has quit working."

"That's all I have, Sergeant."

The whole time Bill was dressing, the Doc asked questions, and as Bill buttoned his last button, he could stand it no longer.

"What is it, Doc?"

"It's one of three things, but we don't have anything like an X-ray, so I can't give you a definitive diagnosis. But I can give you some advice!"

"Yessir?"

"Sergeant, you don't need an infirmary. You need to be in a real hospital."

The doc did not send him back to the barracks. He put him in a bed in the new infirmary with orders to stay there until he, Dr. Tremble, could get a plane for him out to the States. There was nothing more he could do for him at Thule. The nurse would take care of him until the plane arrived, the doctor told him.

That was when Bill began thinking he might die, and he really didn't want to die in the cold and the dark. But then again, he really did want to finish out the year because only then would his required five years and nine months of service be up. They would all be going home soon anyway. He had to admit he was getting anxious, but he had to make his point.

"Doc, I can wait for a flight just like anyone else."

"How about I put you on a list for 'first available flight?'"

"First available" meant very little to Bill. Back in the day, they'd waited forever to board the first available transport, and if he remembered correctly they'd waited forever for that first supply boat to return. He remembered they'd been told that supplies would arrive in a "few weeks," but the "few" stretched into six miserable weeks. The words "first available" had been bandied about too many times. Still, he was on a list and was going home.

After two weeks, Norris stopped in to tell him that a plane was on its way. It arrived early the next morning, and Dr. Tremble let him go home and pack his bag, but he made Sergeant Price go with him. Deck-and-Peck also stopped in, and they carried his belongings as Price helped him walk out of the infirmary. The little entourage began its long walk from the clinic through the base and out onto the runway.

Bill was pretty much doubled over as he walked beside Norris. He'd not only taken all the pills faithfully, but the doc had given him something else to keep him comfortable for the journey. As they walked toward the airstrip, Bill noticed more men joining in, then a few more following behind. In all, they made quite a parade. Sergeant Price didn't say a word until after they rounded the corner of the last barracks, when they all saw the plane waiting on the runway, and there beside the runway stood not only all of the men from the original thirty who were still there, but many, many more—all spiffed up and dressed for the occasion in clean uniforms, braving the cold without heavy furs and standing at attention. As Bill walked by them, each man saluted him and wished him well, and at the very end of the line stood Joe. He, too, was at attention and saluting, just like the rest. But the other Inuit weren't there. Joe was alone.

One of the new men shouted out as he walked past, "Hey, Pascouli! Come back soon! The fishing is great!"

Bill could not have been prouder of those men. Those men had done more and given him far more than he had given them. Price had arranged the farewell, and all these men were showing him real love and respect! It was the best thing that ever happened to him; he soaked in all the love and recognition, and knew he could face whatever came next.

CHAPTER 11

✴

A Call from Chrysler

May 20, 1953. He might as well have died and gone to heaven. Not only was he about to get his promised, final, honorable discharge when he got to Rapid City Air Force Base, but he'd had a send-off from Greenland with all the good wishes anyone could possibly have asked for. The doctors at Eglin had just let him out of the hospital. He was fine now. Whatever it was they took out of his innards made him feel human again and happy to be at Beavers' house, but he could not stay. He was set to fly to Wyoming.

Virginia and Peggy had gone back to Wyoming. Although he had a ticket in his pocket to go see them, something told him he shouldn't plan to stay. Furthermore, before he could go to Wyoming, he had to report to South Dakota to be discharged from service. Some of the good feelings he'd had in Greenland were beginning to fade.

Mr. and Mrs. Beavers treated him as though he were going to break and made him sit down in the living room in spite of his protests that he was okay. He and Beavers had been talking shop, waiting for the news to come on, and Beavers had gone to the kitchen to get two beers, which he put on the coffee table. Bill took out a packet of cigarettes and offered one to his buddy before he remembered that Beavers smoked a pipe! Bill lit his cigarette, and Beavers filled his pipe with new, sweet-smelling tobacco that he lit with the silver lighter Mrs. Beavers gave him last Christmas. It even had his initials on it.

Beavers leaned back in his overstuffed chair and turned the radio dial; the two men settled down to wait for the news to come on. The clock chimed five, and Mrs. Beavers joined them for the News Hour. It was June of 1953, and the president was about to speak.

"Hard to think of Eisenhower as the president after all these years," said Beavers. "Guess he'll always be the General for us old guys."

"I still like Ike!" said Mrs. Beavers, chuckling at the old campaign slogan. Ike had promised during his campaign to end the war in Korea, and now he had his chance. The world was waiting. Beavers placed his pipe ever so perfectly across the rim of the ash tray as though he had something earth-shattering to say.

"We're getting a color television set! You'll have to come see us more often, Bill."

"Well," said Bill. "guess I'll stick with radio a while longer. You know, where I just came from, we were excited just to have short wave!"

President Eisenhower's strong voice came through. He promised the American people that he would always tell them the truth, and he did. He wanted his citizens to know what was really going on, and the nation loved him for it. If nothing else, Americans knew one thing: news was what it was, good or bad.

That day's bad news was that fighting was still heavy in Korea; the good was that the allies had forced a stalemate somewhere near the 38th parallel, which meant the possibility of bringing all parties to the table. They [the allies] and all parties [America's one-time ally the Soviet Union, plus North Korea and China], were working on an armaments agreement. Ike wanted to be clear. This news did not mean a treaty had been signed, but he believed it did mean hope. America would continue the mission given them by the United Nations General Assembly and continue to fight for South Korea's independence.

The President then told the country good night, and the announcer came back on. That was when Bill learned why Joe was the only Inuit present on the day he left Greenland.

In 1953, at about the same time the U.S. Air Force Base in Greenland opened officially, about the same time Bill was waiting for the plane to take him home, the Air Force had forcibly relocated more than a hundred members of the Thule tribe some two hundred miles farther north, to an even more barren area. Bill could only speculate what had happened; he'd had no inkling any such thing was in the works. What he did know, was that Joe and his family could barely survive when they lived at Pituffik, and he could not imagine what would happen to them in an even harsher environment. In Bill's opinion, evicting Joe's family from the only home the Inuit had known for millennia should never have happened, and it should never have happened like that.

"So Virginia's gone to visit her family?" Beavers interrupted his train of thought.

"Well, she plans to, but right now she's back in Wyoming, so I really appreciate you letting an old guy like me hang around. As soon as I get my stitches out, I'll be heading out, too. I can't thank you enough for taking me in. It means a lot."

Mrs. Beavers was on her way back to the kitchen when the phone rang on the little table in the hall, and Mrs. Beavers picked it up. "For you, Bill."

Bill put out his cigarette and got up to answer the phone. Beavers heard Bill say "hello" before he covered the receiver and turned to Beavers to whisper "It's the Commander!" Beavers waited until he heard Bill say "Thank you, Sir," before his curiosity got the better of him.

"Okay. What?"

Bill looked at them, puzzled. "Apparently there's a job opening at Chrysler Corporation, and if I want to be considered for it I have to be in General Mechling's office tomorrow morning 8:00 am sharp for an interview. He says there's competition, but he also thinks I have a good chance. He said I had a good recommendation. Don't ask me. I don't know anything else."

Major General Edward Mechling, the new commander of the Air Force Armament Center, had his headquarters there at Eglin. He had been the Air Force's

Chief of Staff for Armaments under Eisenhower during the War.

Bill woke the next morning talking to himself. "Why Chrysler? Chrysler builds cars. Why me? Why would anyone recommend a noncommissioned officer for an automotive job? Exactly what is this job?" Finally he concluded that whatever it was about, he needed it, and a job back in Michigan might be good for his whole family.

"Take your resume," said Mrs. Beavers.

"I don't have one," said Bill. Mrs. Beavers offered to help him write one, but there wasn't time.

"So what do you have?"

"A scrapbook."

"That will just have to do. You can't pass this up. Come on, what else did he say?" asked Beavers.

"My name came up in some conversation. I've been recommended. Come on! I don't know a soul at Chrysler." Bill seemed not to comprehend what was happening.

"It doesn't matter who! If it's worth a shot, go for it; just don't count on it. Do you have all your enlistment and assignment records—Lackland, North Dakota, Korea, Greenland—all?"

Bill nodded. "And South Dakota and Eglin."

"Test scores?"

"Yes. That's what got me sent to Greenland."

Morning came, and after tossing all night Bill felt as though he was caught up in a time warp, one in which he only had minutes to complete something that would otherwise take days—or weeks. He put on his best trousers and headed out for the General's office. It was easy to find. Bold letters on the door read "Major General Edward P. Mechling, Commander, Air Force Armament Center."

Bill had hardly taken a seat next to two other men when the General's secretary came out with her notebook, which she consulted before she looked around the room and called "Brosco?" She had to be going in alphabetical order. Bill didn't even look like a soldier; he wore civilian clothes. At least, he thought, his pants were pressed. He was a stickler that way.

The General asked a few basic questions about his family and circumstances, then gave him a brief description of the job, which was to "do some electrical engineering around the Chrysler Building in Warren, Michigan, which was a very small town outside of Detroit, and anything else the Chrysler Corporation might need along those lines."

Apparently, the Department of Defense had awarded a very important contract through ABMA, the Army Ballistic Missiles Agency, to the main Chrysler Corporation in Detroit, Michigan. Whoever got this job would be working for Chrysler—not the DoD. Chrysler was at that very moment trying to negotiate with the current tenant in that Warren building, the Navy, to release some of their floor space to the DoD, but Bill's salary would not depend on the DoD getting that space. He would be working for Chrysler, and Chrysler owned 328,000 acres in Warren, so the General doubted that any space issue would ever pose a significant problem.

Then the General asked for Bill's resume. Bill guessed the jig was up and the interview was about to be over.

"I apologize, Sir. All I have with me is my scrapbook." He placed his scrapbook on the desk in front of the General then backed off and stood there stiffly, with his hands behind his back, fingers crossed.

To Bill's surprise, instead of pushing the book to one side and sending him out to wait with the others, the General took his time reading it, thumbing through it page by page. He would occasionally glance up at Bill and ask some question like "do you have security clearance?" Finally, he looked up from what he was reading.

"Do you want this job?" to which Bill enthusiastically said yes.

When the General reached the last page of the scrapbook, he told Bill to step up to his desk. "You will take this job?"

"Most certainly, Sir," Bill said, still not clear what was going on. Then, the General picked up the receiver and dialed his secretary.

"Mrs. McCurren, please tell the other candidates in the waiting room they can go home. We have found our man."

Bill was speechless. The General then gave Bill the short version of what he would be doing in Detroit. Bill missed a lot of what the man said, but there was one thing he couldn't forget, when Bill asked him if he would be working on Chrysler automobiles, this was the General's answer.

"Son, you will be working with Dr. Wernher von Braun, the German engineer who brought the V-2 rocket from Germany at the end of World War II. The V-2 has a range of about 300 miles and can carry a nuclear warhead. Dr. von Braun has plans to improve on it."

The General went on to explain that he had only recently become chief of the special weapons team at Eglin, and—as part of the U.S. Air Force Command in Washington—he'd been asked to find an engineer just like Bill. This was in confidence, of course, as was all information he might pick up at Eglin.

"You've been around missiles at Eglin, right?

"You mean the Tarzan, Sir? Very little. I've been the Ground Power section; I managed diesel engines for the radar, Sir," Bill said. "but I had a little exposure to missiles in Korea, Sir. Also very little. You see, I flew reconnaissance."

"Well, good luck then," he said.

"Thank you, Sir."

It had only been three days since Bill had been discharged from the hospital; he was still wrapped in bandages. His sutures itched. They'd let him out of the hospital after two days, and his whole world had changed on that third day. He now had one week to get his stitches out and report to the Chrysler Defense Plant in Warren, Michigan. What was Virginia going to say?

Mrs. Beavers let him in; Beavers was right beside her. "How did it go?" they asked in unison.

"To tell the truth, I'm not real sure. But I do have a job, and I am going to have to be in Detroit in a week. Sorry, Beavers. We're going to have to put another hold on that fishing trip."

"Congratulations. We can always go fishing. There's an even better spot for hunting not far from Detroit—across Lake Huron in Canada. Hunting and fishing." Then Beavers remembered he needed his car back. "See if your doctor will take your stitches out. I'll take you to the airport."

Bill spent a lot of time on the phone with Virginia that afternoon. Their old house in Wyoming was on the market, but she wanted to keep it. She didn't want to live in Eglin … or Detroit. Bill told her he would re-pack his bag for South Dakota and Wyoming, but he would be there soon. He couldn't stay long because he had to go to South Dakota first. He reminded her that to be officially discharged from the services, he had to report to Rapid City Air Force Base there. Only then would his tour of duty be declared officially over; only then would he be free go go to Wyoming.

Beavers took him to the airport. It hadn't even been a week since his operation, and here he was with a three-legged ticket that began in South Dakota. He took his civilian clothes with him when he reported to headquarters there. He had served his full five years and nine months. After only two days in South Dakota he had been debriefed and received his final discharge papers, but somehow the prospect of going on to Wyoming left him with a load that was heavier than it had been when he left for Eglin.

The flight from South Dakota to Wyoming was short and uneventful. He hadn't expected Virginia to meet him so took a taxi to the house. Virginia was waiting. She had made him a coffee.

"Where's Peggy? Is she already asleep?" He'd hardly put his suitcase down.

"No, she's still next door; Elisabeth will feed her. You know I work late, I'm not usually home until 8:00 o'clock. I had to leave work to get here." Bill didn't know. It seemed late for a baby to wait that long.

They kissed, but they kissed like strangers.

"I've made pancakes for you. It's not much, but I haven't been to the store today."

He couldn't decide whether to hold her close then or wait a bit to see if she would love him first, but she did not. It would take time to get to know one another again. It would be like starting over. So much had changed.

"Sure. Great! The house looks wonderful." And so went the conversation until time to pick Peggy up next door. Virginia went by herself and came back with Peggy sound asleep on her shoulder. Bill held her a few minutes before Virginia put her to bed in her crib.

"I'll sleep on the sofa, if you like," said Bill. And she took him up on it.

The short weekend ended, and Virginia returned to work. Bill took care of Peggy all day, learning about babies the same way he approached laying out an airfield. He had to know the specs, but it was worth the trouble because he found himself doting over Peggy and falling in love with his wife all over again. The rest of the week went like that until, finally, on his very last day in Wyoming, the truth came out. Virginia admitted she was in love with the man at the telephone company, which was why she didn't want to go back to Eglin. Surely, after he'd been gone so much of their married life, he understood. She couldn't bring herself to give up what she had, and he couldn't ask her to.

When time came for Bill to fly on to Detroit, their embrace was sincere. Nothing had really changed except that Bill realized he had long been preparing himself for something like this. Now that it was real, it broke his heart.

Chrysler expected him. He would check himself into a hotel. There would be plenty of time to look for another place later, but it was dark by the time he reached Detroit, and he wanted to hear a familiar voice. He would call Adele, his older sister and the only one of all eight siblings who considered herself his champion. He had to let her, at least, know he was here.

The phone rang, and Adele answered. He could hear children in the background. He had hardly gotten the word out that he was here to go to work for

Chrysler than she told him about an apartment in the area that she thought she'd be able to arrange for him. He hadn't seen her for all those many years, but now that she and her husband were settled in the Upper Peninsula, he promised to come see her. He didn't have time, he said, to talk about family, but he did want her to know that he and Virginia were getting a divorce. He told her where he planned to stay and that he would call her again tomorrow. She told him to hold off going to a hotel and give her an hour; she was sure she could make arrangements for him to go to the apartment.

It surprised him how light he felt after talking to her. Just having someone in his family he could talk to close by was a wonderful thing. Adele was always the brick in the family.

He was exhausted, and decided to go down the street to a bar and order himself some dinner and a glass of Black and White, not just to regain his equilibrium but in honor of the Atwell Four. It was something he had never done: sitting alone in a strange bar drinking Scotch. But there he was. He took his time eating, then asked the restaurant owner where he might find a telephone. The owner let him use his. Bill dialed the old home in Sault Ste. Marie, but no one answered, not even one of the little ones, so he left a message.

By the time he checked with Adele, she had indeed secured a room in the apartment, just as she said she would. He thanked her and promised to call when he settled in, hailed a taxi, and in less than fifteen minutes was at the Cliffs, the one his sister had told him about. Once inside, he walked up the four flights of stairs carrying his luggage, the weight of which helped orient him to his new life. Room 411 had a reasonably sized coat closet, but he wasn't ready to unpack. Tomorrow he would thank Adele again ... and try to reach his more likable brothers, Jack and Walter.

Monday morning found him ready to go, and after another taxi ride he found himself looking up at huge black letters on a white building in Warren, Michigan, that said: "Chrysler Jet Engine Plant." This, he thought, was supposed to be

Chrysle Defense Division or was it? He checked the letter again. This was the address. The building looked exactly like the picture on the flyer. It was an enormous white concrete building with pillars flanking its main door that reminded him of icebergs.

Once inside he asked for directions to the Defense Division. Chrysler's buildings were spread out all over the place, and this one was just part of Chrysler's million-acre complex known by the world as "Chrysler Corporation in Detroit, Michigan." He walked down the corridor past the door on the first floor where the Navy was still building jet engines, then took the elevator up to the second. He was still healing and could only hope this new job would be more mental than physical, at least for a few more weeks. He patted his stomach for good luck.

When he opened the impressive fifteen-foot glass doors, he expected an impressive interior, but what he saw was a facility that had definitely seen better days. He was in the anteroom of someone's office that was empty except for one fellow—and he was about to leave. Bill stopped him.

"Sorry, I'm the new engineer. Am I in the right place? Bill Brosco's the name."

"Nice to meet you. I'm Curry … Howard Curry, but I'm the last one in here. Everyone else is in the auditorium. Come with me."

Apparently, all employees were supposed to attend a lecture that morning. The auditorium was right down the hall. They could hear chatter coming from its open doors.

"I'm looking for a Mr. Keller?" he asked the young lady sitting at a table checking off names.

She pointed at the auditorium. "He's probably in there."

When he told her his name, she checked it off and handed him a folder full of papers. It startled him to think that he was already on a list, but at least the paper proved he was at the right place. He had expected to meet Mr. Keller in an office, but evidently that was not going to happen. Perhaps said office was also in disrepair.

There were at least a dozen other men going in, and when he asked, he was told all were engineers. He could see what he was in for. It occurred to him, though, that this was Day One for him but probably old stuff for the rest. He stretched before he went in, then found a seat among the several dozen men already seated there. They were waiting for the speaker, whom he assumed would be Mr. Keller.

As he looked for a seat, he introduced himself to others who looked as lost as he was, then sat down with Curry. Both were busily studying the packet when the Chief of Ordnance for the United States Army Ballistic Missiles Agency, also known as ABMA, took the podium. He introduced himself, welcomed the engineers, and began his lecture, which was about what Chrysler had been doing during the Cold War. Bill was as ready as he ever would be.

"Anticommunism alone will not stop undemocratic governments from co-ercing their own citizens to go to war against their neighbors. President Truman made it clear when he said 'we will always support democracies against author-itarian threats, especially any nation threatened by Moscow.' If we are to answer this call, America must have a strong military as well as a defense system with a strong-enough reputation to deter any super-power, especially the USSR and North Korea and China, from invading its democratic neighbors. As part of his position, Truman wanted the U.S. to have an Army-directed Ballistic Guided Mis-sile Program, and well ... this is it!" The Chief waited for the clapping to subside.

"And you are all part of it. Our scientists and designers at Redstone Arsenal in Huntsville are hard at it, designing better rockets and weapons. And I know President Eisenhower and the Department of Defense trust Chrysler to build their designs!" A little more clapping before the Chief of Staff continued.

"Mr. K. T. Keller was for many years the president of Chrysler here, but he has a new job. He's been appointed by President Eisenhower to serve as Director of our nation's Office of Guided Missiles. Mr. Keller couldn't make it today, but I promise you will all meet him.

"There is one more man you will meet next week, and that is the director of this same program in Huntsville, Alabama. He is the man we have all heard so much about, Dr. Wernher von Braun. Dr. Von Braun and the men in Huntsville are responsible for all design work regarding the Redstone missile in Huntsville, Alabama. As for Chrysler, Chrysler's Warren Plant here in Detroit will use Dr. von Braun's designs to build them. Now I want to hear a real round of applause, because every major manufacturer in the United States has competed for this contract and Chrysler has won it!"

He waited for the applause to subside. "Are there any questions?"

Bill had plenty: What rocket? What nuclear warheads? Were they all experimental or were they more of the same? Who was going to test them? All Bill knew after all of this was that German V-2 rockets were still being used in White Sands, New Mexico, and were in some way part of a viable rocket-building program right here in Michigan.

The room buzzed as though this was what they'd been waiting for! The answers to their questions came across like bullet points: The U.S. Navy brought Wernher von Braun and many of his scientists out of Peenemunde, Germany. Von Braun and some 1700 other Nazis went to Fort Bliss, Texas, before World War II ended. The Germans in Ft. Bliss were not all scientists, and all German scientists were not in America. Many scientists had ended up in the hands of the Soviet Union. And another name for this Cold War was "deterrence."

The name Wernher von Braun was now on the lips of every engineer in the room. As the auditorium cleared, the speaker stopped Bill and J. Howard Curry, and took them aside.

"Mr. Curry, you've been with us a while, but I'm not sure you've worked here at Warren. I want you and Mr. Brosco to come with me to look at the space where you'll be working. I'm afraid it may not be big enough, but you will be the judge. That's why you're here—to assess our facilities and be sure Chrysler has all the tools it needs to build these rockets. You will have a chance to see Dr. Von Braun later this week. I understand you both have had experience in construction?" He turned to Bill, "Greenland, was it?"

"Yes, Sir. I worked in Thule."

"I understand you're coming in as our plant manager. This Warren Plant may be a temporary one for the job, though, because the Navy insists they're going to keep making jet engines and they need the space, but just between you and me, nobody is buying those jet engines any more. Anyway, Chrysler has 1,200,000 square feet of floor space so we just have to work it out."

Curry said he thought surely the existing space would be enough. That shouldn't be a problem. Then the Chief turned to Bill.

"As Plant Manager, I want you to let me know what you need and what you plan to do to retrofit our building. Keep in mind that Chrysler is committed to building for Dr. von Braun, and you are both part of Dr. von Braun's team. I'll be around later if you have any questions," and he left them alone in the room.

To Bill, that was not enough information, but it would have to do. The whole day left him wondering what he'd gotten himself into this time. The chief had done little to allay either of their fears, and all they had for guidance was a fistful of papers that was supposed to "tell them everything they needed to know about their new boss, Dr. Wernher von Braun." but nothing could have been further from the truth.

CHAPTER 12

❋

Serious Stuff

"Yeah. His plane just landed. He should be here in about thirty minutes. You'll have time to look these papers over again before you go in. The two engineers shuffled through their stacks of paper until they came up with the German rocket scientist's credentials.

"Ph.D. at Berlin University, what am I supposed to say?" said Bill.

"You say 'How do you do,'" said the chief. "I'll leave you to it," and he closed the door.

"Come on," said Curry, as the man left. "Dr. von Braun's a scientist, not an engineer."

"Yeah, but he's an engineer, too. This guy was Adolph Hitler's man!"

"I'm not so sure," said Curry. We learned about him when those first Nazis were brought to the States in '47. But after I finished college, I never heard what became of any of them."

Bill turned to Curry, puzzled. "You have a degree? What am I doing here? I'm not kidding. I haven't even finished high school." Bill could feel perspiration on his forehead.

"I could have sworn you had a Ph.D. and I was the inadequate one! Anyhow, don't worry. They send new guys down the street to Chrysler Institute for Engineering."

Bill let loose. "No way you could know, but I never finished high school, plus

I've been in Greenland most of the last three years and Korea before that. I hadn't been in the States much more a week before I got this job. I don't know any of this stuff: I didn't know Chrysler was in the college business. But even if they send me, it's a helluva long way to a Bachelors."

"You have to take it as it comes."

"I always do, but it's been my experience that you have to work like hell when it does."

They couldn't help wondering about the man Wernher von Braun. Curry figured his leaving Germany probably saved lots of lives. Bill agreed. Curry noted that those seventeen hundred scientists and engineers the Navy brought in were just some of the Germans who came to the States after the war, but they were people Germany no longer wanted. The more Curry talked, the more Bill worried, so he changed the subject and asked Curry what he knew about rockets.

"I've never seen one, but I have seen pictures. The "V" is for vengeance. That's the biggest killing machine the world has ever seen, but just think about all those brains."

"Think of those brains? Think of all the awful stuff that happened in those factories."

"I know, and it has me thinking. Even Oppenheimer asked 'why am I doing this?'"

"I don't like killing, and I don't want to contribute anything else to this world that can do that much damage." Bill conjured up the mushroom cloud that was Hiroshima.

"You can't think that way," said Curry. "This is different. We're doing this to protect ourselves and others; we have to master a lot of technology."

"Yeah. A cold war isn't really a war, is it? It's a standoff; we're supposed to be a superpower to keep other superpowers with weapons of mass destruction from annihilating each other and the whole world. Who called it a Cold War anyway?"

"George Orwell, I think. But he left out the part about hate and deceit." Curry made it very clear that he wasn't at all happy about working with Von Braun.

"This guy knew Hitler—really well—and we're about to shake hands with him!"

Bill bit his lip. He liked Curry; he was a good man, and he was glad to be working alongside him, but for now, he really didn't want Curry to know he had a father named Adolph.

The engineers arrived well before the rocket scientist, and the supervisor escorted them upstairs to get a better look at what they would have to do. First he pointed out a small space cut into a corner on the back side of an enormous room on the second floor in the main building. That would be their office, or workroom, they were welcome to redesign it. Besides the office all they could see were spent engine parts, chains and hooks in the ceiling, and against the wall was nothing but junk. Chrysler had a real mess to fix, and surely this just part of it. Where were the rockets? The ICBMs? Even the tanks that were supposed to be in this building? Then it struck him. That was why he was hired. He was the one who had to fix the Chrysler plant before they could even begin to build such things, and if he'd heard correctly, Dr. von Braun already had a few of those things in his pocket for Chrysler to build.

That night, as he finished unpacking in his new apartment, he was still digesting the fact that he would be working with the German who made the rockets that turned London to rubble. How Chrysler could justify working with a man who'd worked for the enemy he could not understand. Sure the man built rockets, but he built them in underground factories in Peenemunde, Germany, using prisoners as slaves, most of whom had died. Tens of thousands of people from all over Europe died in those factories, mostly Jews but also Poles .

Bill reasoned that perhaps von Braun surrendered to Americans specifically to leave such horrors behind. Perhaps he smuggled those V-2 rockets out of Germany because ... why? And why couldn't the United States build rockets without him?

He and Curry did what they could to clean up the place, but it took a lot more than brooms and dust pans. They had to cut holes in the walls and bring in cranes and huge crushing machines to even make a start. They worked overtime to create enough open space in the plant to welcome Wernher Magnus Maximillian Freiherr von Braun.

On the day Dr. von Braun was to arrive, at least a half dozen engineers from other parts of the Chrysler Plant came early so they could be there when he arrived; all of them wanted a chance to talk to the man, and they were not disappointed.

At first sight, Bill was struck by the height of the man. He was taller than most of the other engineers and held himself board-straight. He had a gentlemanly look about him, his blonde hair groomed and a good tan, thanks to the Arizona sun. His piercing blue eyes meant business, and Bill was probably not the only one motivated by those eyes.

After those in the room had a chance to register their first impressions of von Braun did they notice the man walking beside him. Von Braun did not come alone; he was being shadowed by a heavy-set German with a .45 pistol in his pocket. Von Braun called him "the Major."

"I'm Wernher von Braun," the scientist announced to his audience as he shook each of their hands in succession. "I will be working with all of you, and the Major is here to watch."

"How do you do," said Bill. He couldn't just stand there, but was already out of words.

Curry came to the rescue. "Sir, we've heard so many good things about you, and it is a real honor to have a chance to be even a small part of your Redstone program, and we're looking forward to learning everything we can about it."

After the how-do-you-dos, the Major retired to a far corner of the room where he would remain, silently, for the rest of the afternoon. Most of the other Chrysler employees had their chance to say a word or two to Dr. von Braun before they left, but they were gone now. Only Bill and Curry remained.

Dr. von Braun asked Bill and Curry and some of the other engineers to follow him. He had keys to the next room. There, for the first time, Bill got a peek into the locked room at Chrysler's 200-square-foot floor and at the infamous V-2 rocket lying on that floor, taking up all the space, floor to ceiling.

Von Braun saw Bill scribbling away on his notepad, trying to calculate the sizes of things, and saved him the trouble. His rocket was forty-five feet long, and there were probably about a dozen just like it. He had brought as many as he could from Peenemunde when he surrendered to the Americans, which he explained was the village where Germany built those deadly rockets during World War II. Bill tried not to think about all those citizens in Holland or Belgium or London hearing those German V-2s flying overhead and running for their lives.

"Where are the rest of them?" Curry asked.

"There is one on the floor below us. The rest are in White Sands, New Mexico. Our scientists and other scientists with different backgrounds use them to test their rockets." Bill felt as though the questions were over; they wouldn't be getting any more information.

"You boys," he pointed at Bill and Curry, "bring those boxes over here and open them up … And you, the other engineers," he pointed at some of the stragglers who were on the other side of the room examining the bottom plate of the rocket with its fins, "bring those benches over here and line them up. You'll find all the tools you need in those boxes."

Von Braun watched them scurry until the organizing was done. Then, as the room began to take the required shape, he pointed straight at Bill and Curry.

"I want you to take it apart," and without mincing a word continued, "and I want you to label every single piece as you take it apart and organize the labeled pieces on the benches—starting with that bench." He pointed to the bench still standing against the wall on the far side of the room.

"Yes, Sir. Is there a blueprint, Sir?" asked Bill.

"No. I expect you to make one."

"Yes, Sir!"

Dr. Von Braun looked hard at Bill. "You're no longer in the service, Boy, and you don't have to be so formal. Where are you from?"

"I'm from Michigan, Sir, but my father was from Poland."

"I wondered where you got that blonde hair. Anyway, I'm going to leave you two to it. I will be back in three weeks." And with that, Dr. Von Braun and the Major departed.

Bill and Curry just looked at each other, shrugged, and began searching for blank labels. All thoughts of calling Adele or any brother or sister vanished. He revised his plans again. He would spend his days dismantling the rocket and his nights trying to get his diploma; his high school alma mater would give him his diploma if he passed a few exams.

With enough help, he could disassemble that rocket and still have time to work on the rest of the Warren building. He had to make it ready for whatever came next, which he imagined would one day include monsters even larger than the V-2. He could almost see rockets with huge fuel tanks and bigger engines spread out on all the floors with necessary tools and parts nearby. Then, as he began to take in how big that first V-2 really was, and that the next one would be larger, he wasn't at all sure Chrysler's space could handle it.

Thus Boy Engineer and the Other Guys began at the end where a warhead might go and worked their way backwards, which meant working well into the night and all the next day. They thought they would never finish disassembling the thing but three weeks later, they not only had all 3,000+ pieces labeled and displayed it perfect order, they had a blueprint. Just in time.

Von Braun returned, once more accompanied by the Major, whom Bill and Curry had by now learned was wanted by the British government. The Brits had declared the Major a war criminal and a wanted man. Not only that, the Brits kept tabs on him, tracked him. And when Great Britain and other European nations learned that the wanted man was in Detroit where Chrysler was about to get America's missile program going, the British press turned up everywhere, including Chrysler. The British government would continue trying to extradite him for his war crimes the whole time Bill worked in that building. As for the Major, when he accompanied Dr. von Braun, he just sat quietly in his usual seat in the corner, and the engineers soon forgot he was there.

Bill was always the one who went to the airport to greet Dr. von Braun. He greeted Bill warmly, and seemed more at ease because he always made it a point to pass on a few pleasantries before he would have to give his instructions to them all.

"So, your father was from Poland. I, too, was born in Poland. What town did your father come from?"

"I'm not sure, Sir. He doesn't talk to me about it much. Where did you live in Poland?"

"I was born in a town called Wyrzsk in Pila County there, but of course, the Germans took it over in 1939. They renamed it Wirsitz. So I was born in Poland, but my younger brother was born in Germany."

After they were settled in Detroit and had gone to the Warren building, Curry asked the doctor if his brother had come to the States with him, and indeed he had. His younger brother had been trained in Berlin as a chemical engineer and was now a propulsion expert. His name was Magnus von Braun, and Magnus lived in Huntsville, Alabama.

Bill told Curry about the doctor being born in Poland, and Curry picked up the subject.

"Now that the USSR has taken over Poland, do you know what the Soviets call your town?"

"Of course!" said von Braun. "When the Soviets took it away from the Germans, they called it 'Commune Wyrzysk.' Now, let's see what we have here."

That was perhaps the last of any small talk Bill ever remembered having with Dr. Wernher von Braun. Von Braun walked slowly among the benches, picking up pieces, examining their labels, and putting them back exactly where he found them. Then he walked back to the beginning and began again. Those benches held thousands of parts, and his inspection took several hours. The engineers held their breaths and took down everything he said about each of those 3,000 pieces. In the end, Dr. von Braun took his sharp eyes off those parts and turned his attention to the two engineers still in the room.

"Now, put it back together. I'll see you in three weeks." As he left, he pointed at the blueprint Bill had tacked to the wall and came close to smiling. "And now you can catalog them."

Bill and J. Howard Curry talked over every connection. There were too many choices sometimes, and even when they agreed on what had to happen next, they were afraid they'd make the wrong choice. As they re-assembled the thing there was an opposite worry: when a part wouldn't go back where it should, they may have damaged it.

But by the end of two weeks, they had assembled most of the system, complete with components that fit and some that had suffered a little damage during disassembly and had to be repaired. The various functions now made sense to Bill and Curry; at least sense enough so they could agree upon solutions. If nothing else, they were now working as a team.

By week three, the V-2 lay completed but nonfunctioning on the floor. Even if nothing else pleased Dr. von Braun, the fact that Chrysler's floor had withstood the weight of the monster plus all the heavy tools and men working on it was an accomplishment and a great relief. There was no production line or anything else of the sort that they knew of other than this one at Chrysler, but before heavy construction could begin, Chrysler had shut down the construction project until the engineers could finish their work on the V-2. For a while, the Warren, Michigan, Chrysler building was officially "under construction" or "renovation" depending upon who was talking.

Von Braun arrived on schedule in exactly three weeks and made his way around the supine, reassembled V-2 without a word. He pushed and pulled at mobile parts and had the V-2 turned so he could get a good look at its undersides. When he finished, he turned to Bill.

"I see," said von Braun. "But will it work?'

"The electricals check out. No flaws in the fuel tanks that we can see. Capacitors are good. Sir, all the parts are in good working order!"

"So, you say it can fly?"

"Well, of course—if everything else goes according to plan." That, they explained, was the job for the rest of the team.

For the first time, von Braun exhibited doubt. He pointed straight at Bill and made one more demand. "Show me every step, beginning with the first one, that you would have to take to make it fly."

"It would be better, Sir, if we both were involved."

"You first. I'll ask Mr. Curry for his opinion later."

It occurred to Bill that he should sound more authoritative, so he tried to remember how Mr. Marriott explained things and began with a lot of "if this happens then that should happen." Only then did he realize that every time he used that word should, he cringed. What if it didn't do what it should? There was no escaping it … everything he described, every one of those 3,000 pieces in that monster lying on the floor could fail, and Bill continued to wince every time he used the word should.

"Now," said von Braun, pointing at Curry. "Let's hear what you have to say."

Bill listened intently as J. Howard took his shot at it, thinking surely he would have something to say that would be critical of what Bill said. But it didn't happen.

Later, when Bill and Curry compared notes, they hand-slapped. They'd done the impossible. They'd not only agreed in what they said in their off-the-cuff presentations, but they'd made Dr. Wernher von Braun a happy man. At least, he'd said something to that effect.

By August, using the V-2 as their model and Chrysler's engineers as supervisors, the entire project, including the completed catalog, was done. The Redstone Team would begin by building Redstones in Huntsville, pending the completion and re-opening of Chrysler's Warren plant.

Another of Bill's jobs was to figure out how to move those rockets over land to the testing site, and he arranged for a giant truck-trailer system, made according to the exacting specifications necessary to protect the rocket. Protecting the rocket was the priority. For the time being, they would build and test Redstones in Huntsville. Dr. von Braun's team figured they would have to build twelve Redstone rockets in Huntsville before renovations on the Warren Building could be completed. Once that was done, production could begin in earnest at the Chrysler plant in Detroit.

During all of that August in 1953, Bill and Curry and the other engineers and scientists, along with wives and children, drove from Detroit to Alabama to watch launches and cheer the rockets on as they shot into the air.

Dr. Von Braun was right there every time, applauding along with the rest of his team at Redstone Arsenal. The Chrysler Corporation congratulated the Redstone Team, thanked Bill for the renovation, and turned its attention to fulfilling their contract in Detroit. Everything was fine until it wasn't.

On August 20, all the engineers from Chrysler and Huntsville arrived in Florida just in time to watch the most recent Redstone lift off from Cape Canaveral's LC-4A pad. They were all there, and they all watched silently as one of the engines failed and the rocket fell into the sea.

Dr. Von Braun saw it more as a challenge than a failure. He did not give up; he never did. He would build more Redstone rockets that never would fail again. The Warren Plant at the Chrysler Corporation's facilities in Detroit announced it was open for business. The Redstone prototypes passed their tests, and Chrysler was ready.

CHAPTER 13

Redstone Comes Home

Bill finished. He had redesigned and renovated Chrysler Missile Division's Warren building, and Chrysler was ready to bring Dr. von Braun's Redstone back to Michigan. The great move took place over several weeks, but soon enough, all parts and men were back in Detroit.

If there was anything Bill and Curry learned during those times, it was that Dr. von Braun would never be content to design just one rocket for one conventional purpose. Even though Redstone was the best thing going and Chrysler had orders for more than a hundred Redstones, von Braun always wanted bigger, stronger, brawny rockets and missiles that could cut through Earth's atmosphere and go out into space. He wanted them to orbit around the earth and carry valuable loads into space. Von Braun saw rockets carrying not just nuclear weapons but scientific and human payloads; he dreamt of putting man in space—even sending him to the moon.

In those early days of Redstone, Dr. von Braun designed his rockets in Huntsville and had them built at Chrysler Missile Division's Warren Building. The assembled Redstone would then go to Huntsville to be tested. Chrysler tested components like fuel tanks, nose cones, heat shields, and non-gravity chambers. He especially wanted the engineers at Chrysler to work out how to build a non-gravity chamber so he could respond to all the wild speculations about what might happen to a human being should that human being ever find his or her way into space.

Scientists and other engineers had posed various theories and concerns for years, saying that without gravity a man could go blind or would not be able to pee or far more imaginative disorders. Chrysler was up for it, certain it could perform such a test. The Redstone team's first reasonable design for an anti-gravity chamber was not much more than a huge tube, but it was enough to test all the crazy theories.

The test was simple: they dropped a man in a capsule from a great height down that tube. The man experienced a queasy 5.9 sec-

Chrysler's Redstone Rocket PGM- 11, the first Redstone launched in Huntsville, Alabama in 1953

onds of weightlessness before he was rescued, and the test tube earned the name Vomit Comet. It was a start, but fortunately tests like the Vomit Comet were the exception rather than the rule.

Chrysler's contract called for five Redstones a month, but with all the testing in that small building, the company had been lucky to finish even one on time. The Redstone project needed more space, so Chrysler sent Bill to try begging the Navy for even a few of their 200,000 square feet they were no longer using for jet engines. But even that proved to be not nearly enough. Chrysler needed 400,000 square feet. The engineers figured that should be no problem considering 400,000 square feet was only about one-fourth of all the space in the building.

Bill and Curry went over their figures again and again, and yes they'd figured it all right, 400,000 square feet would definitely do it if the Navy would cooperate. There should be no problem, but there was a problem, a serious political problem. Negotiations involved Washington, D.C.

Bill and Curry were about to have a chance to meet with Mr. K. T. Keller, whom the President of the United States had just made Director of the Army's Ordnance Department. The Navy "owned" the Warren space and had always had full use of whatever it needed of Chrysler's space to build the jet engines, which had been vital in World War II. The Navy still built jet engines there, but not nearly as many as they used to. Business had fallen off as civilian companies began building their own jet engines, but the Navy was not ready to give up their claim to the space, especially not to the Army Ballistic Missiles Agency (ABMA), which was the agency footing the bill for Dr. von Braun's Redstone. When the Department of Defense's Ordnance Department asked to use it, even after Bill had already spent $2,500,000 for the Army to upgrade it, the Navy's answer was a firm no.

This was no Army-Navy game, however. Mr. K. T. Keller had just been appointed to yet another national organization, the newly formed Army Scientific Advisory Panel. He was therefore coming to Detroit in person, as a representative of both the ASAP and the Ordnance Department, to negotiate. Keller had to resolve the conflicting need for space between the Army and its Redstone and the Navy and its jet engines.

Bill met Keller at the airport and brought him back to the Warren Plant. Keller went straight to the auditorium and called a meeting of everyone working on the Redstone.

"Gentlemen," said Keller. "We have not begun to fight, and I am sure we will prevail, but at the moment the Navy is adamant that they need the whole building, that the building's best use is to build jet engines. They argue that if there were an unexpected conflict, there would be an emergency call for jet engines. In fact, they said that even if jet engines would not be needed for a national emergency, they intend to keep the Warren building for some other military purpose."

The Redstone Program was a Cold War project. It considered itself a defensive rather than an aggressive military program. Redstone did not arm their Redstones with warheads, so the Navy continued to argue that they needed their space. They did not deem the Redstone program essential.

The buzz in the room was both worried and confused. How could this happen? It was Chrysler's building and it was Chrysler's job to build Redstones, and everyone was aware that Ordnance was doing what Ordnance did. Still, the Navy didn't have another place for its jet engines. A distressed Keller gave Chrysler's engineers one final blow.

"I have one more thing to add. After all the arguments on both sides have been put forth, the jet engine faction is threatening to mothball the whole building or lease it out to someone else—anyone rather than turn it over to Redstone. We're going to have to shut our program down again, at least until we can straighten this out."

Mr. Keller stayed behind as the room emptied. The only two people left were Bill and Howard Curry. "What about us, Sir?"

He waved around the room. "You're both still Chrysler's project engineers for electrical design, so keep on doing what you're doing. Get the place ready. We're going to build Redstones."

Everyone got the message. In war or peace, and in spite of the setback, between the end of August and the first of December, Chrysler managed to build two more Redstones in a very small space. When the first one crashed into the sea after an attempted launch at the Cape, Wernher von Braun returned to Detroit to walk Bill and Curry through the whole process. Dr. von Braun summed up the Redstone failure this way:

"The thing was too heavy. It could not carry its own load. Chrysler has to build it larger just so it can carry its own load."

He asked questions and they made suggestions as they always did, and the three men talked it out. The conundrum was that for greater thrust, the rocket had to have more fuel, and the more fuel it carried, the heavier the load would be. Bill took von Braun's words to heart, and his thinking went like this: "if the Army wants to keep on building Redstones and if Chrysler is to build them, then it's up to us to figure out how to increase this space, how to double or triple production in the space we've got. If we can do that, we can build them bigger, heavier, and better right here! So that is what we're going to do."

For Bill, every problem or challenge was personal. He began by recalculating and re-examining everything he'd already done in order to look at the allotted space anew—as though even bigger, heavier rockets were already being built there. Every inch had to be tuned to work at capacity, and many of those inches had to do double-duty. Most importantly, the floors had to be reinforced to bear a greater weight.

Chrysler shut down. The larger issue was in the hands of Mr. Keller and the Ordnance Department. Bill did all he could to make good use of his time, sometimes staying to refigure space after Curry and everyone else had gone home, sometimes to work on his diploma. The paper work needed for a veteran to get his diploma turned out to be a correspondence course that he had to complete. It was never enough to count on past experience. Things changed, and learning how to deal with change was where a man's future lay.

He was just finishing another letter to Colorado State to sign up for more courses when the Chief stopped in. He saw a light in the window, he said, and had come to investigate. When Bill told him what he was doing, the Chief offered to take the completed form to the office. He would even have one of Chrysler's officers sign it! The Chief had wanted to talk to him anyway, he said.

"You know about all the construction going on down the street, I'm sure … where Ford's Special Design Division is building a new luxury car construction complex? Well, I understand there is a big problem down at the main building on Ford Motor Company's construction site."

"Yeah, I see the buildings going up. Don't tell me Ford's moving in on the rocket business, too!"

The Chief laughed. "No, not that. Their issue has to do with drinking water. They say their drinking water is contaminated, and they blame us for it! For Pete's sake, it all comes from the Detroit River, and we all treat it on site. I just wanted to let you know that they called me this morning and threatened to shut us down. I've talked them out of any such drastic action for now. My guess is that once their construction crew leaves, their water will clear up. Anyway, I don't think there's anything wrong with Chrysler's water, but you better check it out."

Bill's worksheet was still on the clipboard. The Chief took it off and looked closer at it.

"Project Engineer for Electrical Design?" he looked up from the worksheet. "I thought you had a degree. Why are you applying to Colorado?"

"I don't have degree, Sir. I've only had a few courses, but they were enough for the Air Force to make me their "teaching engineer." I began working on my degree when I was at Fort Warren, but it all went out the window when I was in Korea and Greenland. But no, I don't have a degree. It's time I got one."

"Ok, there's something we can do about that," he said as he jotted something on a blank sheet in the clipboard. He handed it back to Bill.

"Here! As soon as you have time, take this down the road to the Chrysler Institute of Engineering building and register there for a course or two. It comes with the job, and I hear they have great instructors. No need to stop anything you're doing; just get in your hours and you'll finish. Good luck, fella."

Bill couldn't believe his ears. As Sergeant Price used to say, he was happier than a kid who just found out he had two peckers. He would spend his days in the Warren Building cleaning up jet engine carcasses and nights at the Institute of Engineering with other engineers.

That night, he went home to his little apartment, trudged up several flights of stairs and pulled out a can of Spam and made batter for a few waffles. He'd grown accustomed to the limited menu, which didn't differ widely from what they served up on shovels in Greenland. He was feeling very lonely when someone knocked on his door.

He opened it with a scowl on his face, ready to let whoever it was have it for knocking on his door this late at night, but when he saw who it was, he was a little surprised. There, in the hall, was the girl next door, whom he'd only met once. Her name, if he remembered correctly, was Beverly, and she worked at General Motors in a building just down from the Ford Plant.

"I am so sorry. I didn't mean to be so gruff. It's just that it's been a long day," said Bill. He wasn't about to invite her in. Anyway, she was a mess, dripping wet and not even wearing a raincoat. She was almost in tears. Bill could never stand it when a woman cried.

"I've been out on the street looking for my dog," she said. "He ran out of the apartment barking, then down the stairs, and that's the last time I saw him. I'm afraid someone may have left the door open for a split second and he slipped outside in the rain."

"How long ago did he go missing?"

"Oh, maybe fifteen minutes. I'm really not sure. Can you help me find him?"

Bill hadn't seen any dogs when he opened the door to the street, but it was possible a dog that small could have slipped by him and he not notice. Anyway, he would never turn down a chance to help a woman in distress.

"Glad to," he said and turned off the burner on the stove.

"Oh, thank you. His name is 'Baby.'"

He knew the little dachshund. They walked out together, both with umbrellas, but the minute they stepped outside, a strong wind whipped them inside out. The umbrellas were useless, but they continued to call for Baby up and down the alley. They had no luck in the alley so decided to split up to tackle the streets.

Bill went one way and she went the other. To speed things up, he folded his umbrella under his arm, ducked his head and ran through the rain. He thought he saw a dachshund running down the sidewalk on the other side of the street and called to him, but the dog kept running. Bill decided to get ahead of him and run interception. It worked. Bill found Baby cowering against a brick wall, and it was easy enough for Bill to pick up a the little wet dog. He held the shivering animal close and managed to open his umbrella over them both on the way back.

By the time he reached the apartment, the wind had died down. He could see Beverly under her umbrella coming toward him. She collected her little dog, thanked him, and all three took the elevator to the fourth floor. She wouldn't stop thanking him and telling him she was sure she would never have been able to catch her dog without him. She was sorry she had to bother him, but she knew she just wasn't fast enough.

"No problem," said Bill.

As she started to unlock her apartment door, Bill wondered what he should do next. The only thing he could think of was to tell her what he was planning for dinner, that he had fried some Spam and was about to make pancakes. He would be very happy to bring her some if she would like, but she declined. So he returned alone to finish his now very cold meal. He dug into his Spam determined he would invite Beverly out for a real dinner real soon. Probably the same time next week.

A month passed and he hadn't heard a thing, either from Mr. Keller or Colorado. So he finally did take Beverly out to dinner and invited her to go with him to look at old airplanes. He was thinking about buying one, he said. He told her he had met another fellow at Chrysler who was also interested in old planes. His name was Cliff. Cliff and Bill had figured that by combining their resources they might even be able to buy one. She said she didn't know anything about planes, of course, but would be happy to learn.

The next day at lunchtime, Bill was flipping through the paper looking for old Mustangs, probably P-51s, when the Chief showed up. He had just stopped by to report on the progress that ABMA and Mr. T.K. Keller and the U.S. Army's

Ordnance Department were making. They had advertised that they wanted to buy a structure big enough to build rockets, which meant they had money to buy a site big enough to build five Redstones a month. Forty-five companies had responded. The Army had a long list of qualified builders, and they had to visit every one of them. Chrysler's Warren building was the last one on the list.

Bill figured he was not just on hold, he was on ice, at least for the Redstone. During the day, he moved walls and grappling hooks, installed equipment for the crew. Those crews came at a premium because construction never stopped anywhere around Detroit. An entire industrial complex was being built in their neighborhood, not just Ford and General Motors.

He had almost forgotten the issue about the drinking water at Ford. He had tested Chrysler's water at source and in the building, not once but several times, and he had reported that it was okay every time. Consequently, he was dumbfounded when Mr. Keller called Bill in person to tell him that the Ford Company was shutting Chrysler down until something was done about the contamination of Ford's water supply. They were about to be sued.

"Don't worry, Sir." Bill told him. "Our water is fine. There has to be a leak somewhere in their lines, and I will find it."

The next morning when he drove in, he saw a crew at the other end of the parking lot putting up a sign in front of the entrance to their building. He pulled into his usual space and killed the engine. For a weekday morning, something was terribly wrong. There were only two cars in the whole lot. He sat for a minute, hesitant to go in. Then he saw a secretary he recognized. She was coming out of the main door carrying a poster board, which she proceeded to tack onto the side door. Then she turned and walked toward her car. He was about to stop her and ask her what was going on, when he read the sign. It read: "This Plant Is Closed Temporarily Due To Water Contamination."

He didn't have to move; she was headed his way. He rolled down his window, and she came over. Apparently the Ford plant down the street had acquired a court order to shut Chrysler down because they had contaminated Ford's water. Chrysler, according to her, was being sued.

"It's all been tested," he told her. "It's all okay," but she just shook her head and left.

"Some damn lawyer put Ford up to this," said Bill mumbling to himself, "some lawyer who knows nothing about the water system." Chrysler's water like everyone else's was treated before it entered the building. Furthermore, when treated water left the building, it flowed into a ditch dug for that purpose. It had all been tested. It was all okay. Bill's thoughts ran in two directions: first, Ford was wrong and secondly, damned if this wasn't just one more delay in finishing the renovation.

At the rate they were going, they would never be able to build anything more than Redstones, which meant the Jupiter was in trouble—the Jupiter that was supposed to carry the heavier payload, the Jupiter they hoped would go into orbit.

Dr. von Braun would not be happy. He was the genius, the civilian technical head of the Army Ballistic Missile Agency (ABMA) in Huntsville who had already made history with the Redstone. To him, the Redstone was nothing more than a glorified V-2, a distraction. He needed a booster powerful enough to send a capsule large enough to hold a man into orbit, and he had named this one Jupiter.

Bill left his brief case and jacket in Engineering and walked five minutes to the lab. His only option was to test Chrysler's water again, from every angle, both at source and at all distribution points. Before the day was out, he knew there was nothing wrong with Chrysler's water. The problem had to be at Ford.

The real problem now was how on earth he could ever get permission to go inside Ford's chain link fences and begin digging. He had to have an accomplice, and with signs all over Ford's building site saying "do not enter," it was going to be hard to find an accomplice.

That night, over shrimp cocktails at a nice restaurant, he had an idea.

"Beverly, I need your help this time."

"Certainly," she said sweetly.

"I need you to drive me in my car over to the Ford plant later on tonight, as soon as I can go home and get my tools."

"Tonight? The Ford plant? Why? It's surrounded by wire fencing!" He told her the whole story, or at least enough to make his case. Beverly would know where he was going and why, but no more. He left out the part about what he planned to do. If he were caught trespassing he'd pay the price, but he didn't want Beverly involved. He wouldn't tell her how he would get inside or what he would do when he did, but he would tell her that he knew what he was doing.

By 10:30 after all of the night shift had gone home, he and Beverly drove over to the site with a shovel, a bag full of rags and jars, and a few other tools. They parked outside the gate and switched places so that Beverly was in the driver's seat with the key. Beverly looked distraught.

"What am I supposed to do?"

"Just wait here by the gate, please, while I go check something. If anyone comes along, just tell them you thought you had a flat tire … and lock your door."

"Please be careful," she told him.

He would, he assured her, as he pushed the shovel and the three pint jars in the bag through the chain link fence. Then, looking both ways, he climbed that fence, re-bagged the three screw-top pint jars, and carried his load deep into Ford Motor Company's huge yard. It took a while but he finally located the water main and estimated the locations of the valves. Biting his lip and concentrating, he checked out all the gullies and crevices he could find and was back at the car within the hour. He had his answer.

Bill took over the driver's seat and handed all the pint jars, now full of water samples, over to Beverly. She wanted details, but he again refused to tell her any more than she already knew, which was that those jars held water from Ford Motor Company's water lines as well as specimens from gullies taken closer to the building. The source of the contents in those jars would be obvious, he told her.

"Just lay low. I'll explain later," and back he went over the fence to retrieve his equipment.

On his way to work the next day, he dropped the specimens off at the lab. Chrysler's lab techs were on it and promised a full report later that day.

He was waiting for word from the lab when Mr. Keller called from Washington with the news that the Department of Defense had gone through the list and were now asking the government for another $6.5 million to find or build a new facility, ignoring the fact that they already had the Warren building. Mr. Keller added that the Warren building was on the bottom of the list, which he knew already. The DoD's request—and probably Bill's fate—was now in the lap of then Assistant Secretary of Defense, a Mr. Frank Higgins. It was already February and so far they had not found another facility that would fit the requirements for building the Redstone. Chrysler had already spent $2,500,000 on the Warren facility renovations and understood why Mr. Keller was so upset.

Almost as an after-thought, Mr. Keller asked about the contamination. Apparently he hadn't yet heard that the Chrysler plant had been closed, because all he asked was "Have you had a chance to look into our Ford problem?" He said nothing about what might happen to jobs at Chrysler.

"Yes, Sir. I've sent samples of Ford's water to the lab. I'm pretty sure we can prove we're not responsible, and I'll call you as soon as I hear."

"How did you get specimens?" Keller sounded alarmed. "Never mind. I don't want to know."

Bill had his report within the hour. As he had said, it had been an easy diagnosis. Ford's water was indeed contaminated, but not by Chrysler. He thought it wise to discuss his next step with Curry, even though Curry had no responsibility for managing the plant. Curry was a good listener, and when he heard Bill out, the two men agreed it was best if Bill went alone to confront Ford through its own plant manager.

Now came the part for which Bill had prepared himself: a personal visit to Ford Motor Company's Plant Manager, a friendly visit, one plant manager to another trying to solve a problem. He had already told the man that Chrysler's water had tested clean and that he was doing everything he could to get to the bottom of the problem.

So Bill put on his overcoat and walked down the street to the new Ford plant with the damning report from the lab inside his brief case. When the two men confronted each other, Bill held out his hand for the red-faced, angry man to shake. Ford's man ignored the gesture.

The man simply said, "We expect Chrysler to fix this problem ... Now!"

Bill persuaded the man to be kind enough to tour the grounds with him, that he thought he had an answer, so they walked together to the far end of Ford's property where Maintenance was busy working on the chains that pulled automobiles up off the floor. Underneath the conveyor belt section was a deep pit that they used to catch wastewater after they hosed off the floors. It smelled of sewage, and Bill turned on his flashlight and bent down to shine it in the back of that pit where water spilled in from outside through a grate. He signaled the Ford manager to take a look.

There were four or five pairs of eyes looking back at them, shining through the dark. There in the recesses of Ford's water system was an entire family of raccoons that had found a safe place to stay. They had even gnawed their way through the grating and taken up permanent residence in that pit. Ford's water was full of raccoon poop!

Maintenance confirmed that they used the space under those conveyor belts to flush the system on a regular basis. The problem was that their wash water had backed up into Ford's main water supply. Chrysler was exonerated. He couldn't wait to tell Beverly and Mr. Keller.

As soon as he got back to the plant, he took the sign down and called Mr. Keller. Bill was laughing so hard he could hardly get his story out, and Mr. Keller got a laugh out of it, too. He thanked Bill for all his hard work. The missile program would live to see another day.

That done, Mr. Keller gave him the latest report on Redstone's search for a bigger plant. Forty-four of the forty-five companies that had been in the running to become Redstone's new building site had already had their buildings inspected. The Department of Defense was about to inspect the last company on the list, which everyone knew was Chrysler's own Warren plant. Mr. Higgins from the Department of Defense would arrive the next morning at 8:00 a.m. sharp to assess the Warren Plant.

For the past ten months, Bill and Curry had never stopped working, so there was a lot of construction debris on the floors that had to be gone by morning. Together, they found enough volunteers who would stay up all night with them and clean up. The two engineers then set about organizing themselves and their records. They knew that handing Mr. Higgins a few drawings was not enough.

By September 24, 1954, Ordnance had come and gone. They made their decision, and the jet engine plant had been declared the best place for Chrysler to build rockets, plus it would save the department the already-spent $6.5 million. The decision worked its way up the chain of command and the Assistant Secretary of Defense, Mr. Higgins was there to "encourage" the Navy to let Chrysler use the building for a two-year pilot program. The Navy finally let go.

Mr. Higgens could now announce to the public that of all forty-five contenders for the job of building Redstones, the Chrysler building in Warren, Michigan, had won back their contract. They had been promised the building at least for now. Bill and Curry were exhausted.

Dr. von Braun was furious. The Army-Navy thing had cost him ten months. The fact that he'd had a chance to work on his Jupiter, and that a dozen Redstones had been born in Huntsville was little consolation. He wanted more. He wanted to go into space. He knew his friend and fellow scientist, the famous physicist Dr. James Van Allen who once used von Braun's old V-2 rockets at White Sands to send experiments into the air, wanted him to go into space.

❋

CHAPTER 14

Jupiter AM-18 and Miss Baker

D r. von Braun loved talking to Bill about the times Dr. James Van Allen would come visit him at White Sands to do his own rocket experiments. Mostly, his tests involved commercial balloons to create the lift for an old V-2. Allen called his contraptions "rockoons." He was still trying to figure out how to light one.

According to Dr. von Braun, that same Dr. Van Allen was part of a larger group of scientists from many disciplines and many nations who were working on a global peace program, to be called the International Geophysical Year or IGY, not to be confused with the previous International Polar Years that had taken place in the 1930s and 1940s, during which scientists studied polar regions. Among the many studies proposed for this IGY, sending an animal into space was one.

One cold winter day, Bill met a man from Holloman Air Force Base who worked for the U.S. Aero Med Field Laboratory, which was part of Holloman's Missile Development Center. He had come to Detroit to talk about the possibility of sending an animal into space because Holloman had both mice and monkeys should they need them. Bill and the man were walking together toward the test site for the Jupiter, talking about the animals.

"Mice are no problem," said the man. "You can buy lab rats any time. Monkeys, of course, are more valuable."

"I'm curious. What kind of monkeys? I know you wouldn't have the big ones, the apes and such."

It was true, said the man. Holloman had acquired two monkeys that the scientists were considering—one a squirrel monkey from South America and the other a rhesus from the Islands.

"Would you like to meet them?"

"I sure would!" said Bill.

The two men walked out to the truck where two tiny baby monkeys were sucking on bottles, just like real babies.

"This one is the rhesus, we call her Able. Able will grow up to be about twice the size of Baker here. Baker was born in our lab back in '57. Would you like to hold it?"

"Is it a boy or a girl?" asked Bill.

"Oh, it's a girl."

"Then how do you do, Miss Baker?" he asked as he picked up the tiny thing. Miss Baker clung to his shirt and settled down in his arms. She weighed less than a pound, and he held her close like that for a while, talking to her, telling her how tiny she was. Then he gave "Miss Able" a turn before he had to return both monkeys to the man from Holloman AFB.

Miss Baker, the South American squirrel monkey who would fly on Jupiter AM-15 at Cape Canaveral on May 28, 1959, and return to earth alive

"Has Dr. von Braun seen them? I'll be talking to him after lunch, and I know he will want rats or mice or something before he sends a man into space."

"Not yet," said the man.

Bill wished him luck and returned to the auditorium where Drs. von Braun and Allen were discussing the whole IGY thing. They both wanted to send Jupiter off into space, but for different reasons. Dr. Allen wanted a booster to send a science experiment into space.

Dr. von Braun had to choose whether to use his Jupiter for a peaceful, scientific mission for the IGY or continue developing it as a weapon of war. Until he knew which way he had to go, he could not even begin to think about sending a living creature into space.

Wernher had worked on the Jupiter the whole ten months Chrysler was shut down for "renovations," thanks to General Medaris and ABMA's money, Army funds were available and General Medaris used them to encourage von Braun to keep on working the whole time. Army funds were already allocated to Chrysler as part of their contract, and Medaris was among those who thought the military side of things was being neglected. He was afraid the United States might be doing itself harm. He had long urged Wernher in that direction.

When Bill arrived and mentioned that it might be time to send a monkey into space, that it sounded like a great thing for the Redstone team to do, he must have said the wrong thing. It triggered something else in Dr. von Braun. Anger. Dr. von Braun was furious that the Department of Defense had stopped work at the Warren Building, that the Department of Defense had stopped his work; it had cost him ten months.

"If Hitler had given me enough money, there would be no Redstone!"

Bill couldn't believe he said that, but then again, he'd seen Dr. von Braun angry before. He told Curry about it the next time he saw him.

"Just between us, those jet engines are obsolete. Or at least they're no longer the only kid on the block," said Curry, "I don't think he'll be angry for long."

He was right. The doctor loved churning out proposals, so when he told them he would be back in two weeks, he also told Bill it might not be a bad idea after all to let one of Holloman's monkeys ride on a Jupiter. Then he gave the engineers just enough time to clean up before the next project, and when he said that, Bill knew the Jupiter was just a beginning.

"Sir, will we be able to keep on working after our two years are up?" The Navy had only agreed to let Chrysler continue using their space for two years. They'd called the Redstone program Chrysler's "Pilot Project," which didn't set too well with Chrysler.

Dr. Wernher von Braun answered circuitously by saying the Redstone would complete its mission, pilot project or no pilot project. But when the Doctor declared that the name on the building, which read Chrysler Jet Engine Plant and Naval Industrial U.S. Army Reserve Aircraft Plant, would have to change, that was an answer.

Bill and Curry upped the ante and worked even harder into finish the building, and all were rewarded when Mr. Higgins announced that he was confident they could now think of the old jet engine plant as Redstone's Manufacturing Program's permanent home. A new Chief of Ordnance, Major General Elbert L. Ford signed a whole new contract with Chrysler, one that made Chrysler the prime contractor and turned the pilot program into a permanent one. It was then that William R. Brosco, Chrysler's Plant Manager, was authorized to put up a new sign on the building. When their sign read Chrysler Corporation Missile Division, it ended two years of squabbling with the stroke of a pen.

As far as Major General Ford was concerned, Chrysler's contract was just one of dozens he had to sign to keep the Redstone moving. The others included American Rocketdyne, General Motors, Sperry Rand, and even Ford, but Chrysler was still the prime contractor, and Chrysler's engineers were still the ones who had to bring all pieces together.

Thus reassured, Bill and Curry began working on the rest of the 400,000 square feet with an eye to multiplying the number of Redstones that could be built. That 400,000 square feet felt like Mammoth Cave, and like the first 200,000 square feet, the work space was littered with cast-off jet engine parts and other debris. It was definitely not the smart-looking, intricately machined, logical and humming place one might expect rockets-in-waiting to be born. Rather, it looked more like a hand-me-down with remnants of braces, brackets, and tools that would be of no use for building the kind of future rockets that Von Braun had in mind. Anything larger than a Redstone, including the Jupiter, would still be a problem.

By 1959, it was time to send Jupiter C to Cape Canaveral to be tested. When Jupiter C arrived, the Cape renamed it "Jupiter AM-18" to indicate the number of tests already completed at Cape Canaveral. This one, the eighteenth to be tested, would have two little monkeys on board.

Jupiter arrived at the Cape, but it still had to be assembled. They used cranes and wenches to set it upright in the vertical assembly facility before they could even begin. There was much to be done before they could even consider moving Jupiter to its launch pad.

Bill and Curry were among the crowd at Cape Canaveral. Hundreds of people had come to Florida to watch the test, but the audience had to wait. Bill had several cups of coffee and was walking toward the launch pad when he saw his friend from the lab at Holloman AFB.

"Hey there. Remember me? I met you in Detroit when you brought your little monkeys with you." The man recognized him immediately, and Bill asked about the two little monkeys. "If I remember correctly, they should be about two years old. Are they here?"

They were not only there, said the man, they were both going up into space. They were Jupiter's payload, healthy and well trained to perform certain tasks upon signals. He was sure they would come back alive, but it was too late for Bill to say goodbye. Miss Able and Miss Baker were already comfortably strapped inside the capsule on top of the Jupiter.

The vehicle launched successfully. Bill watched, but all he could think of was all the stress, the G loads, the weightlessness, the 300-mile climb straight up, the incredible speed and horrific deceleration those two little monkeys would have to experience. The Redstone would accelerate to 10,000 miles an hour very soon.

Wernher was talking on about the latest version of Redstone, which had also been given a name change. Its name was now "Juno," which indicated how even the smallest design change could change a name. Dr. von Braun planned to build a Juno II, then a Juno III and a Juno IV, but when he got to Juno V, von Braun said no. Juno V was different. It would have to be an infant Saturn, and a Saturn V would do far more than just orbit. Saturn V would send a man into space. It would be part of the new Project Apollo, which would be the first real vehicle in space.

Jupiter AM-18 detached, and Bill's attention turned back to Miss Baker. The little monkeys had soon entered outer space. The craft reached 300 miles straight up before it started back to earth and crashed into the sea. The whole flight took only sixteen minutes. Bill waited for news.

Both monkeys arrived alive, but sadly Miss Able didn't make it. She died during her rescue when they anesthetized her in order to remove the metal leads implanted in her body. Miss Baker survived without injuries and in good health. From that day on, Bill would call Miss Baker the "first woman in space."

After all the crazy politics and threats from Ford, Bill figured it was time to make amends to Beverly. He would take a day off, go home, and shave; maybe press his shirt, then take her out somewhere nice. He called.

"General Motors," said the operator and put him through.

"I've missed seeing you, Bev. Any chance you might like to go out to dinner?"

"It really has been a while! Sure. You can fill me in on what's going on at Chrysler, too."

"Yeah. You probably know more than I do." Bill no sooner hung up than Beavers called.

"Hey, Bill! Whaddaya say about a trip to Canada in the Fall? The hunting's great," Beavers suggested. "I just heard about a place up in Toronto that's the best—and it's easy to get a flight." Bill burst out laughing.

"Hey, Beavers! Maybe I'll buy a plane. Just kidding. My life's not my own right now, but maybe next Fall?"

That night, he took Beverly out to dinner and filled her in on the aftermath of the Ford raccoon story. Watching her laugh made him laugh, too. Then he told her all about Beavers, all the plans they'd made and all the fishing trips they'd never made.

"From all the stories you've told me about Beavers, I think I would like him," she said.

"You would, and you'd like Mrs. Beavers, too. She likes to cook!"

Only after he saw her off at her door did he realize he had done most of the talking that evening. Things were getting a little serious, he thought, at least that's the way it looked to him. But now was no time to be daydreaming about the future. He and Curry had work to do. He expected Dr. von Braun to arrive on the two o'clock Eastern flight from Huntsville. They definitely had work to do.

Dr. von Braun arrived on time and seemed pleased to see the now somewhat cleaner jet engine facility, but he wasn't happy. It had to do with the whole Warren plant debacle. He was sure he couldn't count on anything until that jet engine plant was gone—not cleaned up. Gone. But, he said, he hadn't come to Detroit to discuss the jet engine plant. He wanted a complete change in direction—a program for a rocket much bigger and heavier that could go into space and return to earth. For a start, Bill and Curry had to build and test a solar furnace for him.

"Sir, I don't know a thing about a solar furnace. Curry … you?"

"Me neither! I'm your chemical engineer." Then he realized Dr. von Braun was looking at them both as though they had told them the world had come to an end.

"Sir, does it have something to do with testing a capsule? You can bet we'll figure it out," said Curry, and Bill breathed a sigh of relief. Curry to the rescue!

Dr. von Braun told them not to worry, he would talk them through it … as soon as his lecture to the engineers was over. To nobody's surprise, his lecture turned out to be about his proposal to Huntsville for an entire space program.

"If we go into space," (which sounded more like "if ve go into space" with a hint of his native German) he paused and began again. "We will go into space, and we have to come back. So we must first build a container that can be used to test a nose cone with a payload as heavy as a man. We must be sure it can survive a return trip to Earth. I have with me a design from the Huntsville team. We will build it, and we must test it after it is built. I am counting on you here at Chrysler to build a container to test it in.

"Your container, your solar furnace, must conform to whatever configuration is necessary. He paused as if to take questions, but when there were none, he continued.

"The most important part of this test is that this solar furnace must withstand the same extreme stresses that any vehicle entering Earth's atmosphere would have to withstand."

Dr. von Braun then laid out the parameters for the heat and pressure a nose cone was likely to suffer as it re-entered Earth's atmosphere and fell to Earth. He laid it out kilogram by kilogram and pound by pound. Curry wrote furiously in his notebook.

"The Huntsville team will make a miniature model of it first; we always test our designs on miniature models. You, however, will be responsible for perfecting and building the actual furnace for us." Bill and Curry nodded.

"No failures, okay? Every part of this tank must pass all tests. Our vehicle will have to fly at least 18,000 miles an hour, and it will encounter huge stresses when it hits the atmosphere. The worst of these stresses is, no doubt, the most lethal one: heat. A returning vehicle must withstand horrific heat—up to 3,000 degrees Fahrenheit. Those are the two most important factors with which we must begin, but there will be many other traps. Do you understand?"

When von Braun finished, he looked at Bill with his sharp blue eyes as though to say "get to it" and it reminded Bill of a bird of prey. Curry and Bill both said yes as did the rest of the engineers. Then the auditorium emptied. When only the three remained behind to look at the sketches, the Doctor laid them out on the table.

"Take a look, then we go over materials." Bill and Curry consulted each other and asked questions as von Braun stood over them, thinking to himself out loud. "We will one day have a live human being in there. We want to fly him to the moon with this thing. We also want him to come back alive and in good shape." The engineers nodded assent at each pronouncement.

"Yes, Sir!" they said in unison. Dr. Von Braun's English and his instructions were now perfectly clear.

The meeting over, he sent Bill and Curry off to see what Chrysler could contribute in the way of space and equipment. He also wanted them back at what had been the jet engine plant with their report within the hour, before he had to catch his plane back to Huntsville.

Thus began tackling their long to-do lists anew and answering questions about materials, inventorying whatever they could find that would prove useful from both obsolete and workable equipment, then another list of what they didn't have. Once the two engineers were as satisfied as they were going to be with their plan, they returned to the room where Dr. von Braun waited. He studied their faces.

"Can you do it?"

Of course they could, they told him. But after he was gone, the two men confessed to each other that this furnace was not only one mighty big, mighty costly experiment, but if it were a success, there were bound to be more such complex engineering experiments in their future. As they began to consider all the elements necessary, Curry wrote them down. The list seemed overwhelming: not only more and larger solar furnaces, anti-gravity chambers, fuel tanks, fin placement configurations, engine placement, exhaust control, Y rings and so on, all of which had to be moved about by myriad forms of equipment. They had no illusions: they were beginning at the beginning.

Chrysler was only one of hundreds of contractors working on Huntsville designs, but the Chrysler Missile Plant was prepared for whatever von Braun chose to throw at it. Bill and Curry began work with conceptual drawings and dimensions, but dug deeper to figure variations in, for instance, heat and pressure. Their biggest drawback was no real nose cone to test. No such payload had yet gone into outer space much less come back to Earth. They were flying blind but took off anyway. Dr. von Braun wanted the project completed within six weeks.

CHAPTER 15

Super-Jupiter

D r. von Braun had also set his sights on converting his intermediate ballistic missile into an intercontinental missile—or at least a medium-range ballistic missile. In other words, a Jupiter that could take out a target a continent away rather than just the target next door. Jupiter was perhaps the first missile heavy enough and strong enough to do the job. He just had to win over the Department of Defense.

Names also mattered to Dr. von Braun. Whenever he came to Detroit, he made it clear that he came representing the "Redstone Ballistic Missile Complex" in Huntsville, Alabama. The name of the Chrysler Defense Plant in Warren, Michigan mattered. What did they think about renaming it the "Chrysler Missile Plant in Highland Park, Michigan?"

The word "missile" was the problem. Because the world was about to enter its first "International Geophysical Year " in January1957, anything that might be construed as a weapon of war would be frowned upon around the globe. Chrysler might be treading on thin ice if they chose to use the word "missile" in their name. The corporate world was, like the government, an actor in the Cold War, which was why Chrysler had heretofore used the word defense to its name. It was a way to fend off critics. Neither Chrysler nor the United States wanted the reputation as a nation only interested in playing the game of "who has the most lethal weapon."

Washington did not want more atomic bombs. Both the United States and the USSR understood people were afraid. The Enola Gay had dropped the atomic bomb on Hiroshima and Nagasaki. The United States had tested the hydrogen bomb in the Marshall Islands, and the USSR had now tested a thermonuclear bomb. Leaders of nations all over the globe were either trying to allay their citizens' fears or were using that fear to their advantage. The IGY was begun in large part as an attempt to do some of the allaying.

The Western world's hero was NATO, the North Atlantic Treaty Organization. West Germany, France, Great Britain, and the rest of the Western allies who came through World War II together, signed that agreement. They formed NATO. Without NATO, member countries would not be able to depend upon each other for the strength in numbers needed to stop another nation from attacking them. The treaty stated that if any country anywhere ever attacked a democratic member country, all members of NATO would come to its defense. Furthermore, a nation under such an attack had every right to retaliate and if justified to use a nuclear weapon to defend itself. The goal was to deter war, at least a war as costly as World War II. The Soviet Union and its Eastern allies formed its own "Eastern Block" and signed its own treaty: the Warsaw Pact. The question now was: is this Cold War a standoff or is it a war? Dr. von Braun's projects depended on the answer.

Hopes ran high for the International Geophysical Year. Dr. von Braun's friend, James van Allen, and scientists from many disciplines and many nations around the world met to tackle that problem back in 1955. By mid-1956, those scientists had their IGY. Scientists and nations who joined in that international effort began to promote any effort that would study and try to solve Earth's questions and problems, not only on land but under the sea and out in space. The National Science Foundation and the National Academy of Sciences took on the responsibilities of setting dates and parameters, and monitoring all projects. For instance, if a nation wished to send rockets into space, at least during the IGY— between January 1, 1957 and December 31, 1957, they must do so for scientific purposes only. The rules forbade sending anything into space to spy on another nation or to start a war, and any beneficial scientific finding made during that IGY had to benefit all nations, not just those participating in the IGY. President Eisenhower firmly believed that the Soviet Union had to be a participant or the whole project would fail. If there would ever be peace, those two biggest nations had to be in it together.

Nations began submitting their plans to the IGY two years before the IGY began. The United States chose to explore outer space, and the Soviet Union chose the same. This was not intended to be a space race but rather a cooperative effort. When 1957 arrived, the IGY began.

Bill turned to the challenges at Chrysler. Dr. Wernher von Braun had added one more scientist to the roll at Chrysler—his brother, Magnus von Braun, who had once worked alongside his brother in Nazi Germany in the underground complex known as Peenemunde. Magnus and his family were on their way to Detroit.

The Huntsville team were designers; the Detroit team were builders. Magnus was a Huntsville man and a chemical engineer whose specialty was rocket propulsion, and Mr. K. T. Keller had promoted him to executive status before he sent Magnus to Detroit to be in charge of the propulsion lab. He was a welcome addition. Bill, as always, was Magnus von Braun's welcoming committee as well as the one to help him settle into his new office.

All this happened while Bill was trying to build a monster tank to test a re-entry vehicle. He met Magnus the day he arrived, but Magnus said he wasn't quite ready to tour the building. He needed time to settle his family into their new home in Detroit. Bill and Curry left him to his devices and didn't expect to see him for a while. So both men were taken by surprise when Magnus showed up just as they began work on the tank. Bill's first impression was that Magnus was shorter than his brother, but other than that he seemed just like him.

Magnus introduced himself, then took his time walking around the tank, which was not yet a tank, looking at it from every angle. By the end of the day, when he left the plant, he told them he was sure he was going to like Detroit and looked forward to working with them. Like his brother, he was there to give lectures to the engineers.

"I look forward to seeing you in the auditorium on Monday," he said. "I enjoy lecturing engineers. It helps me keep abreast of everything." Bill and Curry reiterated that they looked forward to his lectures, and Magnus added: "Confidentially, if Wernher were around, I would lecture him, too ... on the side."

Much earlier, Curry—forever the World War II historian—had brought up his concern that Magnus had been working in Germany when the prisoners from the concentration camps were dying in those underground workshops at Peenemunde. Now that Magnus had left the room, Curry continued his diatribe about the German V-2 factory at Peenemunde.

"He's not like his brother. Dr. Wernher was the primary scientist at Peenemunde, the one responsible for the V-2 design. Magnus worked with the people underground. He was not only raised as a Nazi, he later joined the Nationalist Socialist Party, the Nazi Party. We all know that the Nazis were the ones responsible for killing not only Jews and the disabled but priests and teachers and members of the rival Democratic Socialist Party."

"Well, okay, but he was also a pilot!"

"Yeah, but he flew for the Luftwaffe."

The next time Wernher came to town, which was to test the finalized tank, Dr. von Braun told a story on his brother. Apparently, Magnus had in his youth stolen a brick of platinum, and Wernher hadn't found out about it until years later when Magnus tried to sell it to a jeweler in El Paso. The jeweler ratted on him, and Wernher not only condemned Magnus for his actions, but lectured him and beat the socks off of him.

Bill found himself liking Magnus anyway. It probably had something to do with Magnus being the younger brother of a superstar. So when Magnus came the second time, Bill shook his hand in earnest.

"Good morning, Sir. We're so happy to have another member of Dr. von Braun's team with us! We've been short-changed in the research department up here in Detroit for some time, and now that the Redstone is off and running, we really need a propulsion guy. Huntsville is always busy designing improvements for the Redstone rocket, but we don't know what those designs are until they land in our laps."

"Welcome," said Curry. "We need a chemical engineer."

Magnus had to explain that he was not in Detroit to work on Redstones or any other rocket. He was a businessman. Mr. Keller sent him as a businessman. But of course, he looked forward to being there when they tested the solar furnace.

Bill and Curry showed him to the auditorium where his brother, Wernher, was about to speak. As the engineers filed in, they all wanted to meet Magnus, but they also wanted to talk to Wernher. They always did, but Wernher was rarely in Detroit.

"Sir," one of the engineers asked, "do you plan to become an American citizen?"

"No," Wernher said. "I do not plan to become a citizen of the United States. I am a citizen of the world." When they asked him when they would begin working on missiles again, he replied "when we are ready."

After that lengthy discussion, Dr. Von Braun gave what might have been his longest speech ever to the men at Chrysler. He told them they had the backing of the ABMA as well as Redstone and reassured them that the space program was too important to stop; that Chrysler had been chosen as the best company to do the work, that Chrysler had the contract, and the program would continue. He then proceeded to lecture the engineers on space flight, especially space flight that included a human being.

"If we do go into space, can we get back?" asked another engineer.

When Wernher von Braun gave his answer, he wasn't talking to the whole gang—he was talking straight at Bill and Curry.

"Men, we have drawings for a nose cone. Bill and Curry here will use a fuel tank to build a solar furnace large enough to test a real nose cone. You Chrysler men are doing important work. If there will ever be a re-entry vehicle, it and everything in it—human or experimental—must survive when it re-enters Earth's atmosphere. This is a much bigger tank than anything we've built so far, and we will test it here, as soon as it is built." Then he launched into the talk he gave Bill and Curry about max G forces and max pressures and such, ending with "this payload must survive. One day a human being will be that payload, and we want that person to come back both alive and in good shape."

Their curiosity at its height, every engineer in that room wanted to know more about this "re-entry vehicle," but von Braun stopped them there and ushered them all out.

"We'll let you know when we finish."

Later that day, with parts of the experimental tank lying about, Dr. Von Braun brought Magnus back to the "tank" room to go over details with him.

The next day, Bill was taking Dr. von Braun back to the airport as he always did, when it hit him. There was no Major! The Major was missing, and he dared not ask. The Major, always a mystery, would remain just that, a mystery.

Bill was none too happy that all of Chrysler not only knew about their test now but they all wanted to see it and began hanging around. What could he do? He had his orders, and he had to keep doing what he was doing. But he knew that every time one thing was finished, then blam, another one would be right behind it, and so would the audience.

Chrysler employees had no problem borrowing or lending tools from each other. Dr. Von Braun drew the line at borrowing and lending ideas. Visitors were no longer welcome.

Their time was up. Two weeks had passed, and the day for tank testing had arrived. Bill made sure all parts and seams were intact and working before both scientists arrived. Then, when the two von Brauns did arrive, they had to wait... for hours.

It was critical for conditions to be right. They needed more energy than they had ever even contemplated before, and to get that energy, they had to wait for Chrysler to shut down for the night. They needed all parts of the Chrysler plant to close so no other branch anywhere would be drawing electricity. The amount of energy they needed was enough to simulate the most horrific heat a re-entry might encounter. To have all the energy possible depended on Chrysler not using any. They dared not try it until all the lights were off.

Magnus gave up at some point and went home. He had seen enough.

With the thought that no amount of preparation would ever be enough, Bill—the Project Manager for Electrical Design—ran his hand over the giant carcass checking angles, joints, and all visible parts. Chrysler's people had cleaned up the debris, and both men combed the area twice over to be sure no sharp-edged trash was lying around, just in case.

As he looked at the tank, he realized it probably cost more than any nose cone. By now Chrysler had spent far more than the original two million on the renovation, and even more money would have to be spent to re-configure the plant every time a design change came through. This "solar furnace" was a very expensive and very important step both for the Redstone and for Dr. von Braun's Jupiter.

Dark came. Dr. von Braun and the two engineers watched lights go out all over the complex and the city. Then they waited another half-hour to be sure all personnel had gone home. Von Braun, even after his confidant speech, wondered aloud if, even with full use of all of Chrysler's electricity, they would have enough.

There were always questions about enough: enough electricity, enough money, enough planning, enough scientists, enough engineers. The one thing there was never enough of was research and development. There was never enough R&D, and there was no point in asking. If R&D were a pie, the re-entry vehicle would be a minute sliver, probably too small to see. So anything could happen.

Curry pulled the switch. and the room heated up along with the solar furnace. Time seemed to move at a snail's pace as the needle moved on the pressure gauge, slowly, and they sweated while they waited.

Then came the explosion, and all of Detroit felt the blast. That explosion shot rivets in every direction and engineers ducked for their lives and ran for the lead shields they'd built just in case. Lights went out all over Detroit.

"Damn, men!" said Von Braun as the shrapnel stopped flying.

"That's it."

"That's what?"

"That's what happens. Things happen. But I want to thank you for what you've already done before I tell you that I expect you to build another one, one that won't blow up. And … I want you to be ready to test the next one in thirty days."

Bill could not believe what he was hearing. The man didn't yell at them or say "you dummies" or anything like that. He just said "thank you and try again." When the place quieted down and they could see, they examined the holes in the walls made by the shooting rivets. Those holes were bigger than any bullet could make. It was a miracle no one was hurt.

Then, as he put on his hat in preparation for his departure, Dr. von Braun stopped in the doorway. He had to tell them one thing: when they tested the next one, he would be watching it on television, from the safety of his office in Huntsville. Curry and Bill assured him they would be ready in thirty days.

The Chrysler Missile Plant was used to getting calls from Huntsville just about every day after that disaster. Huntsville wanted to know when the next test vehicle would be ready. If the solar furnace made it through the next test, they said, they would send their trucks to come get it and take it to Redstone.

Dr. von Braun installed in-house television in every division in both cities in preparation for the next test, and Bill replaced rivets with bolts, among other things. The next tank withstood the pressure. It was a success, and everyone, in Huntsville and in Michigan, watched those tests from the safety of their television sets.

CHAPTER 16

The IGY and Jupiter C

By now the Redstone had become the most popular launch vehicle for our government and countries all over the world. Air bases around the globe, especially those in NATO countries, continued stockpiling them, and right there in Huntsville, the 40th Field Artillery Missile Group was training on Redstones. The Redstone had earned its nickname: Old Reliable, and Chrysler was rolling out Old Reliables as fast as they could. The old Warren building at Chrysler could comfortably turn out two a month, but five? On top of that, not only was Huntsville sending ever larger and more projects to the Warren plant, but Dr. von Braun wanted to finish building his "brawny booster," the one that could send a rocket all the way to the moon, the one he dubbed Jupiter. That one was already under way and had been even before von Braun proposed it to President Eisenhower as a project for the IGY. When finished, it would be the perfect booster, an intermediate ballistic missile (IMBM) that would be able to provide the 1,500,000 pounds of thrust needed to lift loads of more than 22,000 pounds.

Bill was there scrambling to create enough space when Magnus came in and came up with a possible solution. Why not take one Jupiter, surround it with eight tanks, tie everything all together, then when all eight tanks are full of fuel—alcohol and liquid oxygen—shoot the thing off? Magnus came up with the idea, and he called his eight-tank configuration clustering.

When Chrysler began testing the clustering, the noise on all floors reached epic proportions, and there was no escape. There was nowhere in the building one could avoid it. Only when one shift ended and another one was not yet under way was there a little respite.

Bill began shutting himself in his office for short periods so he could think. He wanted to get the banging and creaking out of his head even if just for a minute before he went back out onto the floor. As for Dr. von Braun, he hardly ever came to Detroit now. He didn't have to, now that they had in-house television. Besides, as long as his engineers were there, he could handle most Chrysler business remotely.

When von Braun did come to Detroit, everyone at the Chrysler Defense Plant knew it because he always took the same Eastern Airlines flight and would always arrive precisely when the men were returning from lunch, so someone would see him coming and send out a signal.

"Dr. von Braun approaching! Get out your notebooks!"

The Doctor's personal appearances nearly always meant a new project, usually one he wanted built right away, and sometimes before the last one had been finished. Now that Redstone rockets were turning out in volume, new designs arrived every day, and every new design meant more conferences in the auditorium. It seemed as though every branch of government had a seat at the Huntsville Rocket Center table in Detroit. The Air Force, Army, Navy, Department of Defense and all its many branches, plus entities like ABMA (the Army Ballistic Missile Agency) and ARPA (the Department of Defense's Research Wing) were all there, either funding a program or standing its way.

In addition to all the noise from the factory, with so many entities having a presence at Chrysler, Bill found himself in the middle of all the push-me pull-you between civilian and military companies and agencies.

One day, Bill made a decision. He took his hat off its hook and waited near the door. When the buzzer sounded and the shift changed, Chrysler's Plant Manager put on his hat and disappeared into the crowd of workers walking together out into the fresh Michigan air for a shift change. Bill walked briskly as though he had no time for small talk, then he veered left and headed toward the automobile side of the Chrysler Corporation. He needed that walk, and, now that the Jupiter was more or less on its own, he would buy himself a decent car. Beverly would like that.

As far as he was concerned, if there was no plane like a Mustang F-15 there was no automobile like a Chrysler Imperial. But which model? Chrysler had finally joined the V8 crowd and announced its latest automobile, one of Chrysler Corporation's "muscle cars," the latest of the Imperial line. The Imperial had always been prized the world over, but the 1957 model had just come on the market, and it came with a raised deck 392-Firepower Hemi. Bill tried to tell himself he need not rush into buying a 1957, but then ... he felt lighter already ...he rounded the corner and entered the front door of Chrysler's Automotive Department.

He took his time wandering among the shiny new cars, trying out the seats, starting the engines. By mid-afternoon, he drove his brand-new, aquamarine blue Imperial off the lot and back to his office. He parked at the far end of the huge parking area at the plant, thinking that fewer cars on that end of the lot might possibly mean fewer scratches. He wanted Beverly to see it in all its new-car glory, so he was content to walk the long way back to his office and into the Warren building.

But the first thing he saw when he entered the building was Curry. Curry was waiting for him at the top of the stairs, wringing his hands.

"The Secretary of Defense, Mr. Charles Wilson, just called. He's sent us a memo. I understand the DoD has sent the same memo to Dr. von Braun in Huntsville and to Magnus, so Dr. von Braun probably has it by now.

"It's awful! They've ordered Chrysler to stop all work on the Jupiter! The DoD will no longer support any rocket that can fly farther than 200 miles."

"You're kidding," said an incredulous Bill. "We've already tested the Jupiter; it can fly 3,000 miles. What the heck? ABMA's paying for it. Does General Medaris know?"

"I'm sure he does," said Curry. "We're up a creek."

"It's the IGY," Bill thought.

Dr. von Braun took the news badly. As a matter of fact, he was furious. If other rockets could be used for peaceful purposes, Dr. von Braun thought, why not his Jupiter? He called Chrysler, and Bill picked up the receiver.

"We're not going to take this, Boy. I'm going to see President Eisenhower himself. Tomorrow. Keep on doing what you're doing."

"You mean working on the Jupiter, Sir?"

"I said, keep on doing what you're doing," and he hung up. All Bill's joy and excitement over buying a Chrysler Imperial went out the window.

Dr. von Braun did go see Ike. He told the President that although his Jupiter Program was in an advanced stage of development as a military weapon, he knew it could be adapted for research purposes. He told the President that he already had plans for a new "Jupiter C," specifically to be a contribution to the IGY. He believed that neither the Redstone nor the Vanguard—like the one proposed by the Navy—could do what a Jupiter C could do.

The President promised Dr. von Braun that he would give it some thought. Dr. von Braun returned to Huntsville.

President Eisenhower knew the carnage in Europe well. He was there. He and his generals fought in that war; they were there when it ended. They watched Stalin and the USSR fail the territories they were supposed to protect and abuse their own citizens after the war. Then, when thousands of Polish citizens and East Germans began starving to death under Soviet control, America began a "Berlin airlift" and flew planeload after planeload of food from its own protectorate, West Germany, into East Germany.

Eisenhower had a story to tell. After Hitler died and it was possible for him to go inside Germany and begin assessing the damage, Eisenhower and several other American generals went to Berlin. Their mission was to do everything in their power to restore that country, and they began by taking young German soldiers as attaches along with them wherever they went. They saw the piles of the dead and dying, toured the horrors in the death camps, assessed the overwhelming destruction in German cities.

As they went, they did what they could to help survivors, reunite families, and set in motion all the wheels they could think of to rebuild that devastated country. As they were about to leave, one of his generals asked one of the young attaches: "Now what do you think of America?" The young soldier curled his lips and replied: "We hate you."

So Eisenhower knew quite well how important arms, and especially nuclear arms, were to American security, but he also knew he had to do something to end that hate. That was why he supported everything about the International Geophysical Year. He wanted it to succeed.

By the beginning of the next week, Eisenhower had an answer and gave it to Dr. von Braun, and von Braun informed all parties concerned. The President had charged Dr. J. R. Killian [not to be confused with Mr. Keller] of the IGY to address the IGY's entire peace mission. Killian alone was now responsible for setting the boundaries for a communications satellite that could go into orbit in outer space, and Killian alone would keep the world informed about all future outer space projects. The President expected all scientists in the midst of such projects to keep Mr. Killian as well as the President informed.

Dr. von Braun now had a pathway for a Jupiter C, but continued working on his own military vehicle, the one he knew would be quite capable of going into space. With General Medaris's blessing and using ARPA money, von Braun developed what he now called his "Super-Jupiter."

The Jupiter C had its own life. Eisenhower responded to Dr. von Braun's request and encouraged him, still under the umbrella of the Army, represented by the DoD's ABMA, to "test fire" his Jupiter C for suborbital flight at Cape

Canaveral. Ike only made one stipulation: that Jupiter C must not carry a military payload.

Jupiter C complied and Dr. Von Braun assured Eisenhower that his payload would be peaceful. As far as the von Braun team was concerned, they decided it best if Dr. von Braun not mention his "Super-Jupiter."

All Jupiters could now accelerate from 9,000 to 18,000 lbs. Rockets being tested elsewhere, by other companies in the U.S., maxed out only at an acceleration of 1,400 lbs. The Jupiters could also lift more weight into space. Whether a Jupiter was used for scientific or peaceful purposes, it had to meet the criterion of a 1,500,000-pound thrust. Designs kept changing.

In the midst of all this concentrated effort, the team working on Redstone-Jupiter and Juno suffered a great blow. With no explanation and no forewarning, the DoD took the Jupiter Program away from Dr. von Braun and gave it to the U.S. Air Force. Bill and Curry speculated wildly, as did everyone else, but nothing made sense. They were left hanging.

General Medaris stepped in again. He promised Dr. von Braun he would make it possible for him to continue his work—under the auspices of and with funds from the DoD's own project, the Advanced Research Project Agency (ARPA).

The next time the Doctor called Chrysler, Bill reminded him that as far back as September, Jupiter C had already flown 680 miles. It had already gone into space and he should not worry.

"Don't give up the ship, Sir. I know you will do it."

That was all the encouragement Dr. von Braun needed. He added the Rocketdyne E-1 engine to his rocket and declared it the "mother of all first-stage boosters," threw in other bells and whistles and finished his Super-Jupiter. It was ready for anything.

As for the IGY, it would never have been successful had Stalin not died back in '53. Stalin had no dreams for his people and brought nothing to his country but

Marshall Space Flight Center Director, Wernher von Braun, presents General J. B. Medaris with a new golf bag. General Medaris (left) was a commander of the Army Ballistic Missile Agency (ABMA) in Redstone Arsenal, AL during 1955-1938. Photo taken January 1959.

famine and death. Under Stalin's authoritarian regime, it would never have been possible for the USSR to participate in a worldwide program like the IGY.

Premier Nikita Khruschev succeeded Stalin, and like President Dwight Eisenhower, Khruschev knew that the success of the IGY depended upon having both superpowers participate. Both leaders were also very aware that whoever won this race would have the leg up that would knock out their rival and win the Cold War. This was a great experiment, but it was also a race that could have life or death consequences.

Bill loved being part of it all, except for the noise in his head, which he attributed to all this bouncing around Chrysler, Dr. von Braun, and the whole team was suffering. When Wernher von Braun and Major General Medaris took their two Jupiter Cs out of storage, and the U.S. government announced new plans for

Chrysler Corporation's new name for the Warren facility in January 1957: U.S. Army Michigan Missile Plant, Detroit Ordnance District.

its Vanguard and Explorer 1 programs, Chrysler painted a new name on their Warren building just outside of Detroit.

The now "Michigan Missile Plant, Warren, Michigan—For the Production of Redstone and Jupiter Missiles," on the facade of the old Warren building indicated that all systems were on go. Operations were in high gear, and they were ready. Then the Soviet Union burst once again onto the scene.

On October 4, 1957, a BBC radio receiver at a monitoring station southeast of London picked up a strange, intermittent signal from outer space. As more and more radios picked up a signal, it became obvious that whatever was up there in the sky was orbiting earth. Only then did the USSR tell the world that they, not the United States, had put the first artificial satellite into orbit. They called it "Sputnik" or in the USSR "Vostok I."

The people on earth who were listening to those beeps on their radios had no idea what the thing was. The world soon knew about Sputnik; and it was clear to everyone that the USSR beat the United States into orbit. But it was not clear why. Many earthlings believed Sputnik was something to be feared.

Dr. Killian and other officials in the IGY program were incensed at what the USSR had just done. Not only did the Soviets not inform other participants that they were about to fly the thing, but they had refused to address one issue that overrode all the others the among nations: the potential for negative consequences if one nation overflew another. Americans everywhere were humiliated, and the worst part of that humiliation was that the whole world saw Sputnik happen live on TV!

America was not only taken by surprise, it was horrified. Even President Eisenhower, who had been told by his CIA that the Soviet Space Program was nothing to worry about, never suspected anything until they were told that the 62-pound Sputnik had left Earth's atmosphere and was already orbiting our planet every 46 minutes.

The world waited for Eisenhower to say something, which he did as soon as he heard about it. He tried to calm his listeners, tried to tell them that everything was going to be okay. He told them that the United States would soon launch its own satellite. He expected Vanguard to go into space on October 23 … and everyone knew Vanguard was bigger than Sputnik.

The von Braun team, including Chrysler and General Medaris at ABMA, went into a huddle. The Space Race was on, and much of the military world in the United States had long ago discussed the Vanguard program and was concerned that the Vanguard would never make it. It had failed test after test—unlike the Jupiter and Redstone that had both passed all their tests. When it was clear that

Dr. Killian had given the Navy's Vanguard rocket the go-ahead, both Dr. Wernher von Braun and Major General Medaris from ABMA threatened to resign. That, however, did not stop the Vanguard from flying.

Chrysler's engineers and all of the von Braun team in Huntsville watched the proceedings on their respective TVs in Michigan and Alabama. The Vanguard rode victoriously toward Cape Canaveral and stood proudly upon its launch pad. It wore "sustainer" engines on its thrust plate and used a "first engine" staging technique. An Atlas rocket lifted Vanguard off the launch pad with all the fire and brimstone that always accompanied such lift-offs, and the rocket shot into the air and blew up. Every single part of it was a failure, both the explosion on the ground and all the rocket's aerodynamic parts. The world dubbed the Vanguard disaster *Kaputnik*.

No sooner had the world gotten used to hearing about the Sputnik than the United States was mortified again when, two weeks later, the Soviet Union launched its second satellite into orbit. "Sputnik 2" was also televised, and the whole world was convinced that American space program was probably over. Americans had to admit that the Soviet Union had indeed won the space race but, they said, not the industrial one.

That night, Bill picked up a few groceries, threw them in the back seat, and set out for the apartment. Lights came on all over the city. It was true, new man-ufacturing plants were popping up every day. Detroit was booming, and it set Bill thinking.

"My life is being run by a committee," he thought. "A very large corporate committee. Or was it a government committee?"

He stopped in to a grocery store. As he put steaks and an expensive wine in his basket, he told himself he was being foolish, extravagant. He wasn't used to such things. But by the time he reached his apartment, he felt even worse. He didn't want to drink all that wine by himself, and he half-heartedly unloaded his potatoes and looked for a skillet in which to cook the steak.

Then came the knock at the door. It was Beverly! Nothing could have brightened him more.

"Are you okay? You just got home?" she asked. "I thought we decided we were going to go out to dinner. We said 5:30, but it's going on eight o'clock! I must have misunderstood … Just wanted to check on you."

"Did I say that? Well, you see, I haven't even unloaded my groceries. What do you say I cook for you tonight?" He added that he also had wine enough for two.

"Oh, Bill, I can only guess. It's your hearing again. The last thing I said was that I wanted us to go out because I didn't want you to have to cook, considering all that's going on. You probably didn't hear me."

That was true. He didn't hear her half the time. It was his damn ears.

"I brought steaks!" he said, trying to be cheerful but finding himself doing just the opposite. "And I have something to show you."

He took her by the elbow and they left together and went downstairs.

"Where are we going?"

"You'll see." Out the door they went and onto the street where the brand new Chrysler Imperial sat waiting for the two of them to take a spin. Taking a spin led to dinner out, and that led to some serious discussion that resulted in returning home to steaks still sitting out on the counter and melted ice cream.

Then, somewhat happier, he went to work the next day and began to read through the stacks of mail and memos that were piling up. That was when he discovered that NATO would no longer use the term "rocket." NATO's world would from now on refer to rockets only as "missiles." This meant paper work if Dr. Von Braun and his ABMA team were to follow suit. It was no surprise that when Redstone became a missile, so did everything else.

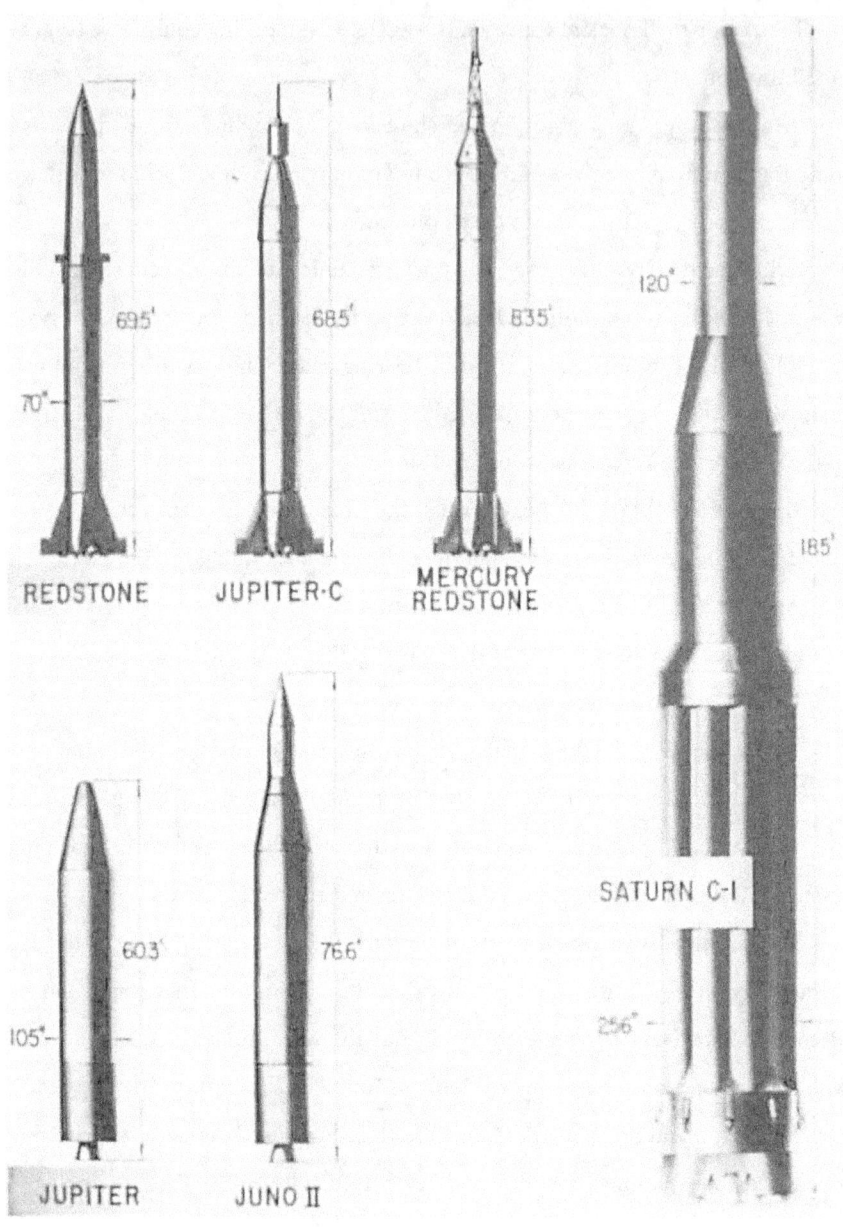

C-1 and earlier vehicles: a. Redstone, b. Jupiter–C, c. Mercury Redstone, d. Jupiter, e. Juno II, and f. Saturn C-1

CHAPTER 17

✱

Michoud and the Saturn Program

No sooner than another "missile design" landed on DoD's desk, did air bases begin asking for both old Redstone missiles and newer Jupiter missiles. Dr. von Braun had just delivered two of them and had orders on the books for forty-five more. Two of NATO's countries were about to receive Jupiters: thirty missiles were to go to Italy and fifteen missiles to Turkey. There was a lot to be said for a name change.

Bill called Dr. von Braun to let him know about all the orders, but it took a minute for the Doctor to respond. His mind had been elsewhere, but he knew one thing.

"Then it's time we started working on the Saturn."

"Saturn, Sir?"

"It's the planet after Jupiter."

If this was going to be a peaceful communications satellite, it was a humdinger. Bill hadn't yet considered how much new stuff like digital machining, heat-resistant alloys, and so on, they might need. How many he didn't know yet, but he did know they would be hard to get.

Chrysler wasn't the only company needing those parts. Companies like the

Rocketdyne Division of North American, Sperry Rand, General Motors and yes, Ford, all had contracts for parts, and Chrysler was not just competing with those other companies for those parts but with itself.

Bill was familiar with many of Chrysler's overseas contractors; he'd been at the Detroit plant not quite five years yet, and Chrysler had sent missiles all over the world to work on their projects in Tokyo, Ireland, Germany, Mexico, France and even Korea. The Chrysler Missile Plant was not only competing with other manufacturers for parts, it was competing with its own Chrysler branches, both within and outside of the United States.

Then came the order that made the most sense. "Salvage what you can."

Tomorrow, thought Bill, we'll begin reconditioning what we have. He was busy making a list when suddenly he remembered. How could he possibly have forgotten his dentist appointment?

He was late, grabbed his hat, and threw on his coat, then hurried to the other side of the lot where he'd parked his new Chrysler. He checked the elongated hood for dings or scratches but there were none.

Dr. Cliff's practice was five miles north of Detroit, but five seemed like fifty that day. By the time he got there, the doctor was waiting. Bill chose Dr. Cliff to be his dentist because the man had an airfield right next to his practice, and they both loved talking airplane. Dr. Cliff did most of the talking; he had to. It was hard to talk with a drill in one's mouth.

"How've you been, Bill?" asked Dr. Cliff as he stuffed hard cardboard bits against the roof of Bill's mouth. "We're going to make X-rays today." He paused only long enough to change his drill bits. "Did you take a look at that Cessna out there? It's my new plane. We can talk about it when I finish." Bill nn-hnnned him.

"It's an old one, but it's got a great engine and I've redone all the controls. Are you still looking for a plane?" Bill tried to nod. Dr. Cliff continued. "I've been flying it back and forth between my farm in Arkansas and work here in Michigan. You know about my farm in Arkansas, don't you? I've cleared an airstrip there, and by the way thanks for all the tips about building airstrips. You know I still have the Stenson, and I want to sell it.," and he laughed. "You are my captive audience, you know."

Bill nodded again, and when the bib came off and after he'd shot the pink stuff into the basin, Bill looked seriously at his dentist and told him yes he would be interested.

"Then come on, let's go! You can take it up."

"I'm not going out on your plane by myself. You are going with me."

"You betcha."

Bill flew that Stenson up and over the St. Mary's River and down and across Mackinaw Island where Adele lived, then back across the water to look for his old home in Sault Ste. Marie, the round island, and the lighthouse. When he saw that lighthouse still standing, he circled it twice, then returned the plane and Dr. Cliff to the landing spot next to the dentist's office.

"Give me some time to think about it," Bill said to Dr. Cliff. "But don't sell it to anyone else until after you've talked to me." Bill left the dentist's office with a huge grin on his face.

He had a lot to talk about with Beverly. A new car and a plane? He doubted she would approve, but then again, he did want her opinion. If he were honest, he didn't want just her opinion, he wanted her approval. Anyway, it might not be a good idea to bring it up just now. He was late again. The Space Race now influenced just about every decision Chrysler and other manufacturers made, which meant Bill's job kept on changing.

Dozens of manufacturers were vying for recognition and contracts which was why, on July 29, 1958, the President of the United States created yet another organization to deal with space projects. The "National Aeronautics and Space Administration" was the latest agency directed to make decisions. NASA controlled whether or not a project would be recognized and funded, and NASA held the fate of the Jupiter program, and all other projects having to do with landing a man on the moon, in its hands.

After the Vanguard disaster, NASA approved the Explorer to be the first American satellite to go into space. Explorer 1 was a satellite designed by the Jet Propulsion Laboratory, the JPL, in California.

Explorer I was a success: it was orbiting space by January 31, 1958, well before the International Geophysical Year ended. Dr. von Braun could not have been happier. Not only had the program used his Jupiter C as the first stage rocket to launch the Explorer, but his friend, Dr. James van Allen had sent his cosmic ray detector along with the other instruments on the satellite. His experiment was designed to study radiation, and when the news came back from space that there was indeed a radiation belt around the world, Dr. van Allen was ecstatic. He had predicted it. To Dr. van Allen and the rest of the world, Explorer 1 was a success.

One of NASA's next directives went first to the ABMA, who sent it on to Dr. von Braun. "It is time to put Jupiter C into orbit," it said. They wanted it in orbit in sixty days. Dr. Wernher von Braun relayed it to Bill, and Bill relayed the order to Curry.

"Sixty days? Damnit, Curry. We could have done that long ago. We've had Jupiter for two years!"

"Well, Bill," said Curry. "Dust one off. I'll give the boss a call."

Curry came back after his next long conversation with Dr. von Braun. "He knows. He's already made plans. He's already been counting them, and the one he's going to send into orbit is actually our 50th Jupiter!"

After a few weeks of preparation and all the ordeal of shipping it to the Cape, that Jupiter shot into the air and performed perfectly, but when NASA called for a Jupiter C to become part of the second Explorer shot, that Jupiter C failed to ignite. It was embarrassing. Everything had to go back to the drawing board for Explorer-friendly modifications.

That was when Dr. von Braun had had enough. There were too many other things he wanted to do rather than keep on modifying the Jupiter C. He declared he would no longer work for the Huntsville group or for NASA's Marshall Space Flight Center (MSFC)—which was the latest name for the Redstone Arsenal. He would resign unless he could finish building that giant booster he had always wanted. He told them again he knew it would send a man into space, but they needed more space to build it.

This presented a real problem for everyone, including Chrysler and NASA. The Huntsville group certainly did not have enough space to build it, and neither did NASA: NASA's headquarters were in the old Dolly Madison building in Washington, D.C.

Then, for the program to grow into what Dr. von Braun envisioned, NASA bought an old 6,000,000 square-foot building east of the city of New Orleans that they called "the Michoud plant" or, as the New Orleanians said it, Michowd plant. NASA bought it for its Marshall Space Flight Center's new Space Program because the MSFC would do whatever it took to keep Dr. von Braun.

That old plant had produced tanks during the war and had plenty of room for more. All the MSFC had to do was renovate it and turn it into an up-to-date, modern facility capable of building the largest conceivable spacecraft that would indeed take humans into orbit and beyond. Never mind it was old; never mind it was far older than World War I. It had once been a sugar plantation in the 1700s.

The somebody who Chrysler decided had to convert that old Michoud plant into a facility that could handle giants like space vehicles, Bill soon learned, was William R. Brosco.

Bill's first reaction was that he had to talk to Beverly. She was his sounding board and friend, if nothing else. He knocked on her door, and she let him in. Then he let loose all the things worrying him, the possibility of having to pick up and move again, the overwhelming logistics that would be involved in renovating such a building, and the very real possibility that Chrysler would become obsolete and he wouldn't have a job at all. The building did not belong to Chrysler; it belonged to the U.S. Government.

Beverly listened wide-eyed. The requirements were definitely staggering. Chrysler had so far only told Bill that if they were going to revamp and fortify that old building, they would want him to do it. Only later did they give him all the details.

The thing dated back to the 1700s and had to be in terrible disrepair. It had always been used for monster-sized vehicles—not only old World War II Sherman tanks but also Howard Hughes's famous Spruce Goose … the biggest wooden airplane ever built. That was how old that building was. Chrysler asked Bill to consider it.

Just being old was not the worst part of that project. The worst part was how much space had to be not only recovered but redesigned. If Bill took the job, he would have to fortify and retrofit forty-three acres under one roof, with at least one quarter of that building or 2,000,000 square feet designated not for the Jupiter but to the brand new Saturn missile program. The Saturn program was so big that it had already been divided up into two parts: Chrysler would build Saturns I, II, III, and IV, and Boeing would build the biggest Saturn of all: Saturn V. Hence Boeing would have control over sixty-five percent of the space or 6,000,000 square feet, and Chrysler would have control of the other thirty-five percent. Both companies needed separate warehouses and office spaces.

"That's quite a challenge, Bill," said Beverly. What else could she say?

After dessert, Bill thought he'd probably been rude and apologized to Beverly for doing all the talking. He hadn't even asked her about her day, but he would make it up to her.

She just changed the subject and reminded him that President Eisenhower was scheduled to speak to the nation that night. He was going to send a Christmas message from outer space to all of America! His voice would be broadcast from the first communications satellite in space. Bill thanked her and turned on the television. How could he forget an event like that? Then, as soon as they settled and the President began to speak, they heard the first voice from space come in over the airwaves:

> *This is the President of the United States speaking. Through the marvels of scientific advance, my voice is coming to you from a satellite circling in outer space. My message is a simple one: Through this unique means I convey to you and to all mankind, America's wish for peace on Earth and goodwill toward men everywhere.*
> — *Press release from the White House, December 19, 1958.*

When the broadcast ended, Bill and Beverly turned their attention back to New Orleans and the Michoud facility and how on earth Bill could renovate a building that had to produce, at the very least, a booster with 1,500,000 pounds of thrust at takeoff.

It was a concern, but it was also an opportunity. He would never say what he was thinking out loud, but taking on a project like renovating a building that was as big as thirteen football fields would bring on a whole new level of stress, and he wasn't at all sure whether he could handle that much stress without Beverly around.

Beverly asked him, quietly, if there were no missiles, what it was that he wanted? He had trouble answering. It was as though he had run out of steam.

"I don't know. What can I do?" He sounded so forlorn that Beverly's heart went out to him.

"We … well, we could go to New Orleans," she said.

"We?" He heard her right, and his problems began to seem minor with her beside him.

The next day, Bill discovered that Chrysler really didn't have a plan. They'd more or less put him in reserve, he thought, because for now they told him, just keep on keeping up the Warren plant enough to handle the work they now had. But Bill had no doubt that keeping on in the Warren building was not the answer. Chrysler had to have a building tall enough and large enough to handle that Saturn program. But where would he fit in until Saturn got off the ground? All he could do was wait. Months passed, but the wait ended.

It ended when Dr. Von Braun called and asked, "What do you know about that building Chrysler's leasing in Louisiana? The government owns it and Chrysler still has the lease on it, but it's just sitting there gathering dust."

"Yessir, there's a Sherman tank manufacturer who's been down there keeping it up, as I understand it," Bill said. He had seen it once in the old days when he was driving over to New Orleans from Eglin. He'd been sight-seeing that day, and someone had directed him to the old plantation. He reassured Dr. von Braun.

"It's plenty big enough to build Saturns. It's probably big enough to build all the Saturns at once, even the Saturn V—the one that will take a man to the moon. The indoor space isn't tall enough to allow rockets, I mean missiles, to stand upright though." Von Braun just listened, but when Bill reached the end of his assessment, he made his declaration.

"We know you can fix it. And, assuming you take the job, we want you and your wife there by the first of the year." Bill accepted the challenge; he would overhaul the Michoud plant beginning in 1961. But he didn't have a wife. Perhaps it was time to marry Beverly.

January 4, 1961. On that Wednesday, Mr. and Mrs. Brosco folded up their tents and moved away from the hills of Michigan to the lowlands and swamps of the South. Beverly found life much harder than she had expected and swamps were not her idea of a good place to live. Bill assured her that although some property around New Orleans was some eight feet below sea level and there would undoubtedly be brackish water on all sides, he was sure there was plenty of land at a slightly higher elevation. They set out to find a house both close enough to the Michoud Ordnance Plant and close enough to Florida's Gulf Coast to drive either place in an hour or so. She was going to love the white sands of the Emerald Coast; he knew she would.

To Bill, all of it, the prospect of marriage and children and sunshine and fishing smacked of paradise anywhere they might find to live. But as he began asking around, several New Orleanians in the area steered them to the Northshore, to Slidell. There, they told him, he would be able to find a lovely spot in an up-and-coming residential area in St. Tammany Parish with plenty of acreage upon which to build a home. Slidell had a population of not much more than 6,000 people, and as they said in Louisiana, Slidell had *boo coo*—the Cajun version of the old French words *beaucoup de*—available land. Living in Louisiana was about to be a whole new experience.

They found a lovely little house in Slidell already completed with enough land around it for child to play in and enough to have a garden, if not a farm. Bill would learn to live with, maybe even love swamp critters as much as he had loved the hares and foxes in Greenland. There were plenty of snakes, alligators, green turtles and other such creatures of nature to keep life interesting. With a little patience, he believed Beverly would come to appreciate them, too. The house was on a country road, a fine place to raise a family.

The day they closed the sale, Bill set off for the St. Tammany Parish court-house with the closing documents in hand to file at the courthouse. January in Louisiana came with both sunshine and pleasantly cool breezes, and Bill breathed in deeply as he parked the car and made his way to the entrance. He opened the heavy doors, and when his eyes adjusted to the dark, he was looking down a long hallway with offices along either side, each of which was behind a Dutch door, most of which were closed. A few were open for business with their top doors open and their bottom ones hooked. The first such office he came upon had a live person inside. It was the Clerk of Court's office, the one he was looking for, or so he thought.

He saw a man working away at his desk, so he sounded the little bell and waited. The man, who wore a rough brown coat and yellow tie, finally laid down his pen and came grumblingly over to the window. He had a scowl on his face and was trying to look official. Bill introduced himself as a new homeowner and asked him where he should go to file his deed.

"You the Yank just moved in?" The man moved in close and pushed his glasses up on his nose to get a better look at the stranger.

"Guess I am. I'm Bill Brosco. Nice to meet you."

Bill was a little startled to be on the man's radar already but reached across the closed panel to shake the man's hand. The official ignored it. After looking Bill over and letting his eyeglasses slip back down on his nose, he jerked his head toward the far end of the hall toward another opaque glass door that also had Clerk of Court written on it in big black letters.

"Down there," the man said and nodded his head in that direction.

It would take a while to get used to the ways of the South, Bill thought.

He filed his deed with no problem and headed back to what they called "East New Orleans" where the Marshall Space Flight Center's old Michoud Ordnance Plant waited. He expected New Orleans would be their home for the rest of his career.

CHAPTER 18

Missile and the Grey Goose

As he drove the Interstate toward the plant, he tried to find the port he'd heard about, the one he knew had to be close by. That port would be the first stop for a missile as it began its journey to Florida to be tested: the Michoud Canal to the Pearl River to the Intracoastal Waterway on to the Gulf and around Florida to Cape Canaveral. He knew there was a Michoud Canal, but he couldn't see it; this might mean he was going to have to build a road, if not some kind of dock. Michoud had built tanks, so the infrastructure to ship them had to be there. All this was running through his mind as he approached the building and drove between the two historic pillars in front.

Those pillars dated back to the 1700s and were the last structures remaining from the days of the original sugar plantation. Beyond the drive and past the building, the rest of Michoud's 829 acres stretched into the distance. He had been told he could use at least three hundred of those acres to expand the existing building or build new ones.

Only when he finally came to a stop in front of the building did he realize he'd spent the whole ride calculating and planning something that he hadn't even seen yet. It was time to shift gears and get a good look at what two million square feet under one roof looked like, and when he entered that forty-seven acre struc-ture, he was not disappointed.

He took his time taking in the details along the walls, right side, rear, left side and there on the left side he saw an open door with a single human being standing in that doorway. He held a clipboard and was so intent on recording the items being unloaded from a truck just outside the door that he didn't even notice Bill. The trucker trudged in and out of the building, bringing in and dropping off boxes just inside the door, and nodding at the supervisor or whoever it was whenever he announced the contents of the box. It reminded Bill of his father gathering information from the ships as they passed by heading for the Soo Locks.

"Hellooo there!" Bill called out. He hadn't wanted to interrupt, but then he didn't want to frighten the man either. "I'm the new man from Detroit."

The supervisor didn't respond, and Bill realized he probably couldn't hear a thing if that truck had its engine running, so he walked across Michoud's mostly dormant and empty floor toward the supervisor. He wanted a look out the farthest door to see if there were another road there, one that might lead to the Michoud Canal or give him a glimpse of the Pearl River, and as he crossed that floor, the plan for a space divided in two for Chrysler and Boeing without short-changing either contractor began to take shape, but where and how to construct a space or spaces that would allow rockets to stand up was a real problem. That space would have to be as tall as a five-story building.

Some of the largest pieces of equipment would come from Boeing and Chrysler's existing facilities, but it wouldn't be the first time. Chrysler had once before dismantled an entire foundry, the one in Chicago, and transported all of it—business, equipment, and all—to New Orleans to be re-installed in this same Michoud Ordnance Plant. This would be his first step.

But Bill had a deadline. It was January, and the Army not only expected Chrysler to begin, complete, retrofit, and be ready to ship the first Saturn to the Cape by October, but they also wanted Saturn II to be in progress even as they sent Saturn I off to be tested. It seemed impossible.

He reached the doorway across the way and introduced himself to the man, whose name was Nichols. Nichols said he'd be done in a minute and would show Bill around then, but until he was free, Bill should wander around and check things out for himself, which he did, from corner to corner. He found what he expected—bits of cylinder heads left from 1951 and the Korean conflict, remnants of disabled Sherman and Patton Tanks, all instruments of a war of the past for which the demand was over.

A few old magazines lay about on top of the bins. He picked up an old 1953 "Aviation Week Airport Directory" that still listed Michoud as a tank factory. In the right-hand column under "status," the directory stated quite clearly what Bill was looking at: "military, inactive, no facilities." Bill took out his pen and added "Chrysler's white elephant."

He found another old magazine that displayed a feature article on Korea, but when he flipped through it, he found nothing other than an article entitled Operation Stardust that talked about China's release of prisoners but said little about North Korean prisoners, so he ditched it. Among other miscellany hanging about on Michoud's walls was an old phone that Bill dusted off and put to his ear. To his amazement, he heard a dial tone and thought of all the calls he and Beavers had made, looking for Bob, but of course even if he had Beavers's phone number, he wouldn't be able to make a long distance call, but it was good to see something functioning.

Sweat began to pour down his face. He'd only been in the building about ten minutes. It was January in Louisiana and the huge building was already hot. It was first thing in the morning, and his priorities had already shifted. Before he did anything else as plant manager, he had to make the hundreds of tons of humidity control equipment operational, beginning with turning on the system.

Like the phone, the fans worked! But the thought of hundreds of men working in a two-million-square-foot building with nothing but built-in fans to keep them cool meant his plan had to begin with finding subcontractors to tear out walls and rip up ceilings. He would have to ask Nichols where to find them. As if he knew what Bill was thinking, Nichols had closed the door behind the departing truck and had walked over to Bill's corner.

Before he did anything, Nichols told him, he had to see the space upstairs where Howard Hughes once lived. He wanted him to see it first because he was pretty sure that Hughes's quarters would be the first thing torn out.

Hughes was a genius and a recluse. Hughes was not dead but very much alive, living like the recluse he was, an old man almost in his sixties hiding somewhere else like California. Howard had not only been Michoud's last tenant and sole occupant for years, but his Hughes Aircraft—and Howard himself—had, in their prime, set dozens of flying records. His Spruce Goose was the largest airplane ever built out of wood, and it was so big that a modern-day Boeing would fit under its wing. The Spruce Goose had long ago gone to some museum, but Hughes's old apartment was still there, and he wanted Bill to see it just the way it was. It had not been touched since Hughes left.

Nichols talked on about the man he'd once known as he and Bill climbed the steep stairs to the old apartment. Nichols said Howard Hughes had always been a hypochondriac, but a very rich hypochondriac, one who had inherited millions of dollars from his father, and his father's fortune had come from oil. Howard had become a multi-billionaire in his own right, long before many millionaires even existed. As the two men climbed the stairs, Nichols went on. Hughes was still a celebrity, and he wasn't just a recluse, he was a germaphobe of the first degree.

At the top of the stairs, Nichols warned Bill that nothing had changed in that apartment since Hughes left. Everything was the same, so Bill should watch his step.

"You know we're going to have to tear it out, don't you?" said Bill, thinking this man might still think of Hughes as a living, crucial part of the Michoud.

"Yessir, and it's a shame," he said as he rustled through a basket full of a green material. "I've gotta show you this. Here! Put this on first." The held up a nylon suit that would swallow Bill and told him to pull it on over his clothes. He had one himself.

"So. I'm going to meet the man?" asked Bill.

Nichols laughed. "More or less! You and Hughes have some things in common, you know. He's an engineer, like you."

"And he's a pilot as am I," Bill added.

Before the superintendent opened the door to Hughes's old apartment, though, he pointed out a rack high up on the wall upon which dozens of those green and white plastic suits still hung. Hughes himself had left them on purpose for visitors to wear. Then they opened the door.

Inside that door was another closed door, and between the two doors there were shelves and shelves of Kleenex boxes with tissues poking out and, below those shelves were metal basins attached to the wall, and each basin had separate spigots for hot and cold running water. Hughes never allowed anyone even close to his inner sanctum unless he or she had hands thoroughly washed before they dared touch the door's sterilized handle.

Bill and Nichols mimed the motions of sterilization, then, before he would open the second door, Nichols picked up a hose and turned a blast of air directly at Bill, and it hit him squarely between the eyes. Then he ran his "air bath" up and down Bill's already sanitarily covered body, a process that no self-respecting germ could possibly survive.

"He's probably a communist," mused Nichols.

Bill grimaced but said nothing. He had heard so much of that kind of talk, especially here in the South. McCarthyism was in full swing all over the country, and the term "communist" was the catch-all, indiscriminately used, convenient put-down for anyone who wasn't just like you. It said nothing about Howard Hughes.

The apartment did. It was still a shambles. The two men spent a minute more in room, absorbing the essence of the strange man who once lived there. When they were done, Nichols closed the door behind them both, reverently, as though for the last time.

The tour over, Bill had finished for the day. Workmen would arrive tomorrow, and reconstruction would begin; but today he would shut the building down and go home to Beverly. He would do what he could to help her get settled, then they would spend the rest of the evening quietly. Beverly was expecting a baby.

He crossed his fingers, hoping the road builders' noise was over for the day. They had moved into a house on a dirt road, just off the highway; but a few months ago, St. Tammany Parish decided to widen the highway, and paving had begun. Beverly had heard nothing but hollering workmen and noisy backhoes and dump trucks, some of which parked at the end of their driveway, ever since they'd moved in. He could only hope they would finish quickly. Beverly needed peace and quiet, and she deserved it.

To his dismay, some of the workmen were still there, and it was quitting time. A couple of them were standing over by the dump truck on what should have been their lawn but was now a muddy construction mess. They weren't doing anything, just standing around and smoking, and as Bill drove past them to his driveway, they ground their cigarettes into the mud with their heels and left. The first thing he heard when he opened the door was a pan dropping on the linoleum floor. Beverly had dropped a pot with a lid on it. It didn't have anything in it, but it made a noise.

It was just nerves, Beverly told him. He spent his time calming her that night, but it took some doing.

He had no time to deal with road builders all that week because contractors and their men were everywhere at the Michoud plant, talking with each other about where to put the assembly lines and consulting with Bill as they marked the whole floor with chalk so they could get their crews started. Bill had already marked off his assembly lines on the floor. Those marks ran the length of the building with the exception of a small part cut out for the supervisor's station. An office would come later. The contractors for Chrysler's side had to see the blueprints for the smaller Saturns, the ones numbered I through IV, and Boeing was still working on the blueprints for its side. They had the far heavier Saturn V to worry about, with its associated tanks, and were still in the planning stage, but Bill was satisfied. It was clear now how those two areas would vary in size but parallel each other because between them and from one end of the building to the other lay a wide ribbon of flooring that already had a name: Bill called it the "Saturn Highway."

The rest of that year Bill turned his full attention to the building, retooling, changing out specialized hooks and grapples, and sorting through the equipment and tools Michoud would need in order to build not only the largest missiles in the world but also the tiniest bolt. He was on his own. Dr. von Braun was too busy sending his Jupiter nuclear missiles to Italy and Turkey and never came to Michoud in those days, although he said he planned to come at some point.

Bill had last seen him in Huntsville. There had been a ceremony in Huntsville when they renamed the old Redstone facility to reflect its new mission, which was to explore space. Redstone had been renamed. It was now the "George C. Marshall Space Flight Center," named for General Marshall. President Eisenhower was pleased, not only because he wanted to honor General Marshall, but he saw the name change as a way to put even more emphasis on America's scientific space programs. He was a strong advocate for the Army Ballistic Missile Agency becoming an integral part of NASA, because he believed ABMA could better serve the nation under their guidance. The change in Redstone's name marked NASA's ascendance.

Although NASA made Dr. Wernher von Braun the director of the MSFC, and MSFC was still responsible for Chrysler's contract work, Dr. von Braun told Bill that he wouldn't be coming to the Louisiana plant much anymore. Even though he would not see him as much, he expected Bill and the Michoud plant to keep operating just as they had always done but as one of MSFC's arms. Bill didn't say anything, but he was going to miss his mentor and was pretty sure such a change in the organizational chart was bound to complicate things.

Back home in Slidell, the road gang had apparently made a mistake. When they dug the ditch to widen the road, they must have cut a sewer pipe, because there was the smell of sewage in the air. A few nights later, on Bill's drive home, he clearly saw where mucky water was pouring out of a cut beside the fresh concrete on the new highway and into an open ditch. He then followed that ditch all the way down the road in front of his house to his driveway, where the water seemed to be pooling.

He called a Tammany Parish Council number that night and got the phone number for the contractors. Bill promised Beverly he would call the Department of Transportation and find out who was responsible for district road improvements. From the next day on, he began making calls to all branches of Louisiana's DOT and Slidell's own road department in earnest.

The night came when Beverly told him that the workmen had come that day and cut the ditch even deeper into the bank so that sewer water was now pouring out onto their grass. The workmen left it that way and they left early. The smell was far worse now even inside the house. The whole family felt nauseated from the odor. Morning came, and Bill now made phone calls to explain his predicament to the Tammany Parish councilmen. The man on the other end of that line responded "Yeah, it'll be gone when they're done." Beverly called, too, telling the man she was expecting a baby any day now and was really worried about the possibility of serious diseases from what was, essentially, an open sewer

Both Bill and Beverly did their best to keep the St. Tammany Parish council members informed about what was going on, but very little changed. When the day finally came that no crew at all showed up, and he was sure work had stopped, Bill knew it was time for him to take drastic measures. He would begin by going to talk to a councilman in person. He wanted the councilman to put him on the agenda at the next St. Tammany Parish council meeting; he would take a sketch with him that showed the easiest way to redirect and contain the sewage. The winter meeting had already taken place; the next meeting wouldn't be until late March.

Bill left early for work and made a side trip to the District office to confront the councilman before the man could claim he didn't have time to talk. Bill was on his best behavior and thanked the man for seeing him. He explained that his wife was about to have a baby and that some sewage line must have broken while they were improving Highway 190 because not only could they smell it, but sewage was now backing up into his yard. Worse yet, he told the man, when it rained, he was afraid that sewage would back up to his house. They needed to address the contours of that ditch.

"Whaddaya want me to do about it, Yank?" as though he knew all about Bill. Bill had never seen the man before. The man's tone made him angry, but he refused to lose his temper.

"Sir," Bill said, "if you are not responsible, please send me to someone who is," to which the councilman answered he was the one responsible, and he would talk to the St. Tammany Parish government the next time they met. He would have someone call Bill.

Weeks went by, and as February turned into March, Beverly gave birth to a healthy baby boy, whom they named Michael. With a newborn around, the urgency of the sewage situation became life-threatening. Bill had heard nothing from any councilmen and decided to take his case to Louisiana's sheriff's department.

At the same time, he had to finish the first Saturn—Saturn I—in time to make it to the Cape by October. He had parts to order and things to do so the men could get on with their work, because work on the Chrysler side of Saturn Lane already included the beginnings of Saturn II, which was moving along at a satisfactory clip. Projects related to the Saturn that depended not only on Chrysler but on Chrysler's contracts with other companies. Vital experiments and tests for anti-gravity chambers, heat shields, drone parachutes, engines, monitoring controls, reentry attitude systems, worldwide tacking systems, track boards, and even the Cape Canaveral Control Center itself were now taking place all around the world. Coordinating all of it was now NASA's job, and NASA had chosen the Marshall Space Flight Center in Huntsville to be the agency in charge of choosing contractors and workmen for ongoing projects for both Chrysler and Boeing.

It was getting complicated.

Boeing's work on the other side of Saturn Highway, the huge Saturn V, was also well along, but Boeing, like Chrysler, had to become involved in ancillary things like renovating an old steel mill they could use for thrust rings and fuel tanks that had to be redesigned to at least twice the size of those on Saturn I. Saturn V had to have bigger tanks to hold more fuel in order to boost bigger and heavier rockets. Saturn V needed maximum thrust if it was going to carry a load with a man in it. The heavier the missile, the more fuel it needed. If and when Saturn V were to send a man to the moon, the whole project would be named Apollo, and the last Saturn, the Saturn V, would take on the name of its payload—Apollo. The first Apollo to pass its test would no doubt mean the last Saturn.

As for Bill, everything important on those assembly lines seemed to be working smoothly; but nothing he had done had made a dent in the problem at home. If St. Tammany Parish couldn't fix it, he would definitely have to go to the Louisiana Sheriff's office, and if that failed, he was afraid he would have run out of options.

Fabrication of Saturn SA-1 booster at the Michoud Assembly Facility, 1969

CHAPTER 19

The Cuban Missile Crisis

After the devastation wrought by the atom bomb in World War II, countries around the world believed that the only way they could protect themselves was to develop and stockpile their own nuclear weapons. Nations everywhere began building bigger and always more lethal weapons to prevent another country from attacking them, which was a dangerous way to go. The United States and other countries looked for a way to prevent nuclear war altogether. In the meanwhile, they had to do what they could to keep the peace. They called it détente, a French word meaning "easing of tensions" that had been used back in World War I.

In early 1961, Dr. von Braun had the design for his first Jupiter, and it was to carry a nuclear missile. America had just sworn in its new president, John F. Kennedy, who inherited what his predecessor, Dwight D. Eisenhower, began: the program scientists from around the world had agreed upon, the International Geophysical Year. The nations who committed themselves to this program believed their cooperation would bring about scientific discoveries in every field, and that missiles could be used for peaceful ventures and not only be weapons of war. Among those countries were scientists who wanted to explore space, and to cooperate in doing so. For this to be a viable step toward peace, Russian scientists had to be part of it.

The world breathed a sigh of relief when Nikita Khruschev agreed, then Ike's term came to an end. John F. Kennedy was in office.

The engineers at the Michoud facility had no idea what Kennedy might do during his term as far as missiles were concerned, but they did know that Dr. von Braun had his Jupiter, and Jupiter and its other configurations, like Juno I, was capable of serving in either war or peace. It had been designed to boost a lethal weapon in war, but it could certainly play a part in exploring outer space. No one would have guessed that Jupiter would play a role in the greatest crisis Kennedy and the United States ever faced.

By 1962, Chrysler had produced dozens of Jupiters. Bill watched Jupiters send the little monkeys into space—the first time with Miss Able and Miss Baker, the second time with Ham. Many scientific experiments went along for the ride. Jupiters coming off the production line also went to America's allies in NATO, like Italy and Turkey, in the hope that by arming American air bases overseas the U.S. could help them protect themselves from aggressors. So far, there had been no such attacks. Détente had not yet been tested.

President Kennedy inherited a problem in Cuba that was growing worse. Cuba, America's close neighbor, had been a vacation spot for Americans, especially Cuban-Americans. At the time, some Americans lived in Cuba, and some Cubans lived in the States. When Fidel Castro took power, he not only rid the country of the dictator, Batista, but began getting rid of anything that suggested American influence. He did this by confiscating American-owned properties as well as from American corporations in Cuba and from Cuban-born American citizens.

President Eisenhower broke diplomatic ties with Cuba even before President Kennedy took office. He also placed an embargo on Cuban goods like sugar, cigars, and pipe tobacco, all of which Americans bought. To keep its own economy strong, Cuba had always relied on its ability to sell these things to Americans, so when Ike put this embargo on those particular goods, Americans complained. Those were some of their favorite things.

As Kennedy grappled with the embargo on Cuba as well as a large population of Cuban Americans who had lost all they had owned in Cuba, the displaced Cubans were desperate. Unbeknown to President Kennedy, some of those desperate Cubans got together to figure out how to get their property back and decided to go talk to the CIA. Together, those Cuban Americans and the CIA formed a plan and took their plan to President Kennedy for approval. Kennedy, against his better judgment, gave the CIA permission to help the Cuban Americans.

Now, with permission to seek help from the CIA, a few of those affected Cuban Americans formed what they called the Cuban Democratic Revolutionary Front, which turned into what amounted to an air-borne posse. It was not a large group, but Americans were shocked when they heard that a Cuban-American posse had flown over Cuba and bombed a Cuban air base on the Bay of Pigs, which was located on Cuba's southwest coast. The conflict at the Bay of Pigs was bad enough, but a few days later—to the absolute horror of Kennedy and other Americans—that same posse invaded Cuba by land.

Cuba naturally fought back. The much larger Cuban army then overpowered the intruders and took them all as prisoners. Not only had the Cuban-Americans lost their battle, but Castro now demanded that Kennedy make reparations before he would release his prisoners. Ironically, that happened on the same day that the U.S. Navy Commander, Alan B. Shepard Jr., was about to ride Freedom 7 into space, and the event went a long way to soften the tensions.

On May 5, 1961, Shepard put his special gloves on, gloves with flashlights in the fingers, and boarded a Mercury capsule. A Redstone rocket launched Mercury into space, and Shepard was the first man to leave the earth's atmosphere. Without those gloves and the light they emitted, Shepard might well have not made it at all: that light shone on the instruments that guided him home. When he returned to earth, his flight was celebrated not just because he had gone into space and returned safely, but because it was a sign that man might really make it to the moon. Americans' hopes rose that they might actually beat the USSR in that race.

That event, as glorious as it was, was quickly forgotten in the middle of rising fears associated with the Cuban missile crisis and Cuba's American prisoners. Anything having to do with the Redstone-Mercury or the Apollo project faded quickly into the background, and whether the U.S. was right or wrong by allowing the CIA to help the Cubans, the die was cast. Kennedy had an emergency on his hands and had to figure out what to do.

Then, an already bad situation turned even worse. President Kennedy was told that the pilot of an American U-2 spy plane had filmed Cubans building facilities to house nuclear missiles all across the island, and they were building these bomb stations with Soviet assistance. Not only that, at the same moment, Soviet ships were in the Atlantic steaming toward the island with nuclear arms on board. It was clear to everyone now: the USSR and Cuba were colluding.

Then, in October 1962, another American U-2 flew over Cuba, and Cuba shot it down as it flew over Cuban airspace and captured its pilot. America and all its citizens were on high alert. Cuba lay no more than ninety miles south of Florida, and the United States now faced the greatest threat it had ever faced: nuclear war.

America had never come that close to another world war. The emergency was clear. Everyone tried to tell Kennedy what to do. Some of his most high-ranking officers pressured him to attack and to attack quickly, before those Soviet ships could land in Cuba. Almost no one in the President's cabinet was opposed to attacking. This was Kennedy's predicament.

From earlier intelligence reports, Kennedy was aware that Soviet premier Nikita Khrushchev had promised to send Soviet arms to Cuba. That was earlier in 1962, when Khrushchev saw nothing wrong with sending Soviet medium- and intermediate-range ballistic missiles to Cuba. But Khrushchev had never before sent missiles to Cuba. Khrushchev had, just as the United States had in other countries, established an air base in Cuba, ostensibly to deter war, as an act of détente.

But this was different. Soviet ships were about to enter territorial waters, and they were delivering nuclear-armed missiles to Castro's doorstep. Kennedy knew Castro's installation sites were ready for the deadly missiles and he could not let it happen. But instead of attacking Cuba, Kennedy—being young and relatively inexperienced and hopeful—picked up the phone, turned on his recorder, and called Nikita Khrushchev. For two days the men talked to each other and sent letters back and forth over the teletype laying out proposals of how to prevent a major conflict. They examined every possible action they could take.

One thing was clear; neither man wanted war. Both leaders understood that any nuclear war would destroy both countries; no country could win such a war. They, and only they, had to defuse the situation. Thirteen days into the crisis, Kennedy and Khruschev made their decision: they would work together on a path toward peace, and that path would be acceptable to both countries. On Day 13 of that Cuban Missile Crisis, on 28 October 1962, Kennedy called in his cabinet and described his conversations with Khruschev to them.

"I will tell you now that it was a very sober two days. There was no discourtesy, no loss of tempers, no threats or ultimatums by either side, no advantage or concession was either gained or given, no major decision was either planned or taken. No spectacular progress was either achieved or pretended."

Khruschev sent the same message to the Soviets: "This is radio Moscow. Premier Khrushchev just sent a message to President Kennedy today that said …"

At some point during those many hours of discussion, Khruschev recognized how inexperienced the new president was and told him so, but he also assured the young president that he would not take advantage of that inexperience. Khruschev wanted both of them to come to an agreement, and they did, two days later. As soon as they made their simultaneous announcements, Soviet ships began to turn around and go home.

Even though Khruschev continued to insist that the USSR never intended anything other than helping Cuba build up its defense system, he promised he would not only remove all Soviet presence from Cuba but he would help assure the return of Cuba's prisoners to their homes—on one condition. Khruschev had a problem with Americans having Jupiter missiles at U.S. air bases abroad. Kennedy had to promise Khruschev that he would remove von Braun's Jupiters from American bases in both Italy and Turkey, seventy-five in all. With that promise and within two days, the two leaders together averted disaster.

Bill was at home when he heard the news. He had Baby Mike on his knee and was feeding him just as his father Adolph did with his younger brothers and sisters, probably with all nine baby Broscos on his knee, and they loved it. Adolph would have been there with his grandson Mike, too, had he been well enough.

As his son rested, Bill listened to the rest of the news. Chrysler's Jupiters were on their way home. It seemed to be an expensive waste of U.S. government money, especially if those Jupiters were coming home only to be stockpiled in some government warehouse. He took it as a signal that the Jupiter program was all but over for the time being.

The next day, Bill was back at work taking stock. That old building, Michoud's Ordnance Plant, had never looked so good; it had become an extravagant and effective home for Saturn rockets! What had once been two ugly storage sheds—one for the wood and one for the metal materials Howard Hughes once needed for his Grey Goose—were now offices and well-organized storage areas for the whole complex. There were no cast-off Jupiters in New Orleans.

Bill was deep in thought when he looked up and noticed Mr. Nichols, now Director of Michoud, walking down Saturn Lane with a group of people dressed in their Sunday-go-to-meeting best. It turned out they were all government dignitaries from one office or another. Mr. Nichols was giving them the grand tour through the facility.

Nichols had changed his tune since Bill encountered him that first day. In those early years, during the first stages of renovating the building when Mr. Nichols was showing Bill around, Nichols wouldn't even bother to look inside the sheds.

"Those sheds are so dang ugly I won't even open 'em!" he'd told Bill, and Bill had taken that as an order to demolish them. But he didn't; instead, he revamped them and made them into fine offices and clean storage areas. Bill found himself grinning because Nichols was proudly showing those visitors each and every one of those re-vamped offices and warehouses, probably because they were only a few of the areas fully finished.

The greater mass of that building, with the exception of Saturn Lane, was very much still under construction. Nichols now led his group down the lane into a Saturn world as seen by both Chrysler and Boeing.

Dr. von Braun's goals for his Saturns had been clear. He expected Michoud to produce five generations of missiles plus an assembly facility capable of building four manned space stations, in this order: intercontinental ballistic missiles by 1961, a lunar orbiter by 1965, and the finale, a fifty-man manned space station by 1967.

Michoud was just a hair behind for the lunar orbit, but past the ICBMs and on its way. Von Braun's team had also declared they would launch Saturn I before the end of 1961, which they had done. Bill had worked with Chrysler solidly every day, and Saturn I had finished in less than a year, well ahead of schedule. Saturn I wouldn't go into orbit yet, but it would go to Cape Canaveral for testing, which had meant a whole new set of challenges for Bill.

Using Dr. von Braun's specifications, Bill had to figure out how to transport this expensive, monster-sized but delicate missile a thousand miles by land and by sea to Cape Canaveral, safely. Because Bill was the engineer in charge of that move, Mr. Nichols expected him to explain it to visitors, and while he explained, work had to be put, briefly, on hold.

"NASA chose the Michoud Plant in large part because it was near a river system and we could make use of that river's backwaters. At the time, NASA didn't have a definite plan, at least no detailed plan, for how to get something as huge as an assembled Saturn from Louisiana to the Gulf of Mexico and to the eastern shore of Florida and the testing facility at Cape Canaveral. We've been working on it. We have to put Saturn I on a special trailer just to take it to a barge on the Michoud Slip, which will then take it out into the Michoud Canal. After that stage it will have to be moved by any method possible to get it to the Intracoastal Waterway and on to the Cape," he paused and looked at Bill. "Mr. Brosco here can tell you all about it; it's been an engineering problem all along, and Mr. Brosco is in charge," he said confidently as he turned his entourage over to Bill.

Caught off guard, all Bill could say was that yes, he would build whatever was necessary to send any and all missiles to Cape Canaveral to be tested and missiles would be launched there and yes, he would be happy to explain what he could.

"What you are looking at is a Saturn I lying on its side. The very weight of it can cause it to deform, so we have to turn it at regular intervals so it won't lose its shape. We also have to make sure it maintains its perfectly smooth round through-out the journey and even after it arrives at the Cape," said Bill to the visitors. "It's almost ready to begin its first journey from here to Cape Canaveral, but the whole project has turned out to be far more difficult than any of us could have imagined.

"Saturn will leave the Michoud Slip on a barge, then it will travel down the Pearl River and onto the Intracoastal Waterway, which will allow us to ship it to the Gulf of Mexico and on around the tip of Florida to Cape Canaveral. You will note that we are not using the Tennessee River; we don't want to risk taking Saturn through its locks. Believe me, I've had experience with locks, and I can't think of anything worse than trying to maneuver this giant through those locks. Sending Saturn I to Cape Canaveral is nothing like sending a first child off to school; it's more like sending that child off on a raft down a river full of alligators."

Bill then had to take questions about the dangers, which meant they first had to understand how surprisingly vulnerable this huge Saturn was, how difficult it was to protect it at each step, because each step brought on its own dangers—not just jolts and mechanical issues but the elements. The Saturn, in all its configurations, had to be protected from salt water because the Intracoastal Waterway was all salt water. Chrysler had designed a heavy-duty, specialized, waterproof, heat-regulated coat to wrap the rocket in before the journey could even begin.

At the end of the visit, Bill brought out a piece of this most recent invention, a heavily-cushioned, extra water-tight material developed to protect the missile at every stage of its journey. This was the first time Mr. Nichols had seen the material, so he took his time examining the sample before he let his guests handle it. Then he proudly explained how important even such a seemingly small step was in the life of any missile; that this little piece of material was just one example of the thousands of items that had to be produced before a Saturn could even begin its journey. With that, Mr. Nichols accompanied his fascinated dignitaries to the door and assured them they were now all experts in missile design.

That done, Bill hunkered down to ready the Saturn for its trip to Cape Canaveral, which involved a lot more than just having protective covering. It weighed somewhere between 60 and 80 tons, and Chrysler had ordered a specially-constructed, heavy-duty flatbed trailer strong enough to carry that much weight plus any additional weight like cranes that might be needed to move it from the truck to the barge, which Michoud's personnel named Pegasus. Bill checked every inch of Saturn's route to be sure there were no flaws anywhere from the plant to the slip, in the new road or the dock.

Pegasus was waiting. Its first job was to hold steady as Michoud's men unloaded the precious cargo off the trailer and lifted it onto the barge in the Michoud Slip. Once everything was tied down and protected from the elements, the loaded Pegasus would pull away and follow the Michoud Canal to the Pearl River, after which Saturn I would have to take another barge, the Palaemon, to the Intracoastal Waterway—the Gulf Coast's inland route to the Atlantic—and from there around Florida to Cape Canaveral.

It began. Bill watched and waited, forever biting his lip as the men brought the Saturn I, swathed in its protective coat, out the door and onto the trailer. But they still had to load the fuel tanks, which were almost as big as the Saturn it-self, and the engines. Once everything was loaded onto the trailer, the composite looked like a giant coffin being pulled by a monster truck. Bill ignored all the onlookers who wanted to know if they should be worried that a coffin that big might be a bad omen.

As the truck drove its somber load down the road to the Michoud Slip, a raft of pelicans fluttered out from the swamp into a bright blue sky and formed themselves into their habitual V shape. The crane operators readied themselves to begin the transfer of the entire assembly onto Pegasus without sinking the barge or harming the giant dehumidifier installed on Pegasus. That dehumidifier was essential; they had to keep Saturn I dry. The truck with its loaded flatbed finally came to a stop on the deck of the Michoud Slip, and its giant steel crane locked itself into place, ready to lift Saturn off the flatbed and onto Pegasus.

They'd thought of everything, or so they thought. The crane took its time picking up Saturn I and cautiously swinging it through the air and onto the Peg-asus, which was level with the dock and firmly anchored beside the Michoud Slip. Bill saw the problem first. As the men began transferring the weight from the trailer to Pegasus, that weight began to push the barge deeper and deeper into the water until the deck was no longer level with the barge.

"Pump the ballast! Slow now, lower the Saturn slowly … Pump it out!"

The team pumped furiously, constantly, and the ballast sank into the Michoud Canal. For all the rest of that day, with the crew shouting orders and metal clank-ing against metal, they wrestled with the weight. Only when they finished did they realize that they never even saw an alligator nor did they ever hear critter sounds or anything else after the herons left the swamplands. Even the gulls had been spooked by the commotion, and when the commotion ended and Pegasus began to make its way to the Michoud Canal, there was silence. All that was left was the sickly sweet smell of organic matter and swamp, pure swamp.

Bill had been so tense he almost bit his tongue off. Not until Pegasus sounded

The barge, Pegasus, leaves NASA's Michoud Assembly Facility in New Orleans.

its horn as it rounded the bend toward the Michoud Canal and began its circuitous journey to the Cape did Bill stop biting his lip. He joined the rest of Michoud's crew, who were standing on the slip waving their Saturn I goodbye. All those men—who had silently held their breaths throughout that long day while tall metal cranes swung Saturn I and tanks and such through the air—began to whoop and chatter. Out into the ever-widening slough, the tugs guided Pegasus down the Michoud Canal toward the backwaters of the Pearl River where a different barge, the Palaemon, and other tugs would help it on to the Intracoastal Waterway.

Bill and family drove the blue-green Chrysler out of their driveway and began their the long trip to Cape Canaveral, Florida, to be there when the Saturn arrived, to watch the test. Bill tried to ignore the awful smell coming from Highway 190 as they drove away. He felt foolish. He could pat himself on the back for moving a Saturn, but was helpless when it came to getting anyone to fix that damaged road. It was a little like a punch in the gut, and he swore to himself that he would take care of it once and for all when they got back.

Upper stages awaited them at the Cape, more stages to be loaded atop Saturn I. That could not have been accomplished without the Cape's brand new, vertical assembly building, which NASA built. All this had to happen before they could move Saturn I from assembly out onto Cape Canaveral's LC-34 pad where it would get its new name and fly.

Beverly and Mike had a front row seat inside to watch the test flight. Outside the building, the sun shone brightly and the now Saturn C-1 sat royally on its pad, the first of its kind to take the test. The whole team reveled in the glory. Even little Mike stopped whatever baby thing he was doing and, curious, turned his head in the direction his parents were looking.

On that October 27, 1961, Saturn C-I was on the pad and ready for lift-off, and when the missile flew without a hitch, Saturn C-1 passed the test. The blast-off was perfect! They all saw it. Its only shortcoming? It had nothing on board but a dummy payload. It was a beginning, but only a beginning. Many more boosters, even many with no payload at all, had to be tested before NASA could move on to its first manned flight.

The little family had lobster for dinner and slept well that night before they had to begin their long trip back. Bill tried to settle down, but the last few weeks had been so intense, all he could think about was that there was an Atlas missile waiting to be finished at the Michoud. Even though Chrysler was not building it, he felt responsible; as long as it was in his building, he was responsible. NASA had chosen Atlas to be the booster for intercontinental ballistic missiles (ICBMs), and its manufacturer, Convair, now shared a small corner of the floor with Chrysler and Boeing. Because Saturn I had proved itself, there was no limit to what might be next.

Bill's next responsibility to Chrysler was Saturn II, which was already well on its way to completion on Chrysler's side of Saturn Lane. The Marshall Space Flight Center insisted on naming its new projects according to the alphabet, so they called this one Saturn B. But Saturn B was not designed to carry a huge pay-load, and the engineers at the Marshall Space Flight Center worried they would never get the moon project off the ground. The next stages for that final Saturn

Program existed only on paper. Far too much time had passed. The lunar space-craft might well be ready long before they finished its booster.

It had been a wonderful vacation, really, but as Bill and his family left for Louisiana, he realized his mind was back on the moon, and like Dr. von Braun, he too wanted to see the long-awaited "brawny booster." Boeing was building it, but every step they took toward the Saturn V, which would carry the Apollo to the moon, seemed to be agonizingly slow. Saturn V had to be almost twice the size and weight of Saturn I. Bill agonized, along with Boeing's personnel, over what else he could do to expedite the project; but, he decided, he wasn't Boeing.

The drive was lovely. They opened the windows, and as wafts of orange blossoms in bloom filled the air, he forgot the odor at home.

CHAPTER 20

Trouble in St. Tammany Parish

Having had some time to clear his head, Bill realized he just might have an idea how to help Boeing finish its Saturn V, but the closer he got to home the more he realized that his priorities lay in stopping contaminated water from pouring into his yard. The rainy season had turned the drainage ditch into a disgustingly putrid lake, the sheer quantity of which had caused it to overflow into their yard. As he turned in to the driveway that night, he couldn't believe what he was seeing. The overflowing ditch now looked like a small river, and nasty sewer water was creeping across his lawn and toward his steps. The odor was horrific. He'd smelled sewer smells off and on all his life, but he'd never smelled anything like this. He'd seen the slush coming out of a gash in the road, but he'd never imagined that river of filthy water would pour out of that gap and make its way into his house. The gap was in Tammany Parish's sewer system, and he could not even begin to predict how much suffering that might cause—now or into the future. Whatever was required of him, he would do it.

He hurried Beverly and Mike inside and shut the doors. While Beverly was unpacking, Bill went into the kitchen and, before he started making sandwiches, began to stuff tea towels under the doors to stop the odor from coming in, but as he bent down close to the door jamb where he was trying to block the odor, he became aware of a different smell. He smelled smoke. Evidently Beverly smelled it, too.

"Bill! I thought you stopped smoking!" said Beverly. He had, he told her, but what she was smelling was not a cigarette.

"I smell smoke, too, but it's not me. It's probably somebody burning trash. Don't worry. Go on back and tend to Mike. I'm going outside for a minute and I'll be right back."

He put on his jacket and headed out the front door. His car was still there in the driveway, but the noticed another vehicle, a truck, parked out on the street behind it, almost blocking it. It looked like a Ford truck, but he wasn't sure. The wind had picked up, and he could see an occasional curl of light gray smoke coming from behind his house, blowing down the driveway toward the road. Bill grabbed the flashlight out of his car and walked around toward his back yard, shining the light in all directions until his flashlight picked up a nightmare.

There in front of him was a man with a torch in his hand, setting fire to the dry leaves under the window in back of his house. For one brief second, the man froze like a deer in the headlights, a man in khaki pants but naked to the waist. Bill saw flames out of the corner of his eye, flames licking at the sideboard and taking hold, but he had no time for the fire now. The man had jumped away from the light and grabbed a giant limb lying nearby and was charging right at Bill with the branch pointed at his chest. Bill told him to back off or he would shoot, but the man kept coming, running straight at Bill and stabbing him hard with the heavy limb. The blow sent Bill reeling backwards. He felt pain shoot through his chest but regained his balance and stood his ground.

"Stop where you are, or I'll shoot you in the face. Not only that, but I'll keep on shooting until you're garbage and nobody will recognize you." Blood was oozing from Bill's chest, and he didn't have a gun, but the man was running toward him, toward the truck. Bill had to catch him…

In spite of the pain and the blood, Bill managed to grab hold of his arm and knock him over backward and into the swampy water. As the man scrambled to get up, Bill tried to hit him again, but the man dodged and ran. Bill yelled at him all the way, saying if he ever saw him again he would kill him, but the man made it to his truck.

As Bill heard the truck's engine rev, the only thing he could think of was that anybody who would set fire to a house with people in it was likely to turn his truck back and ram into him or at least into his car, but the man didn't. He chose to leave and headed in the other direction back toward highway 190.

Bill had to let him go. He had a fire to put out.

Bev heard the ruckus and came outside just in time to see the black truck leave, but when Bill saw her standing there watching him put out the fire, it terrified him even more. He begged her to go back inside and call the police, but she would not leave him. Bill smothered the last of the flames with his bloody jacket and determined that the house had only been singed. They had been lucky; it had not caught fire.

When the police arrived, Bill described the man and his truck as best he could then walked with the officers into the back yard where they took pictures of the blackened wall and Bill's injuries. After the officers made sure the fire was out and no one in the family except Bill was harmed, they told Bill to go to the hospital, then wrote up the incident and left.

There was no way any of them could sleep that night. Bill tried, but was soon pacing the floor thinking about Beverly and Mike alone in that house while he was back at the plant. At least for a little while, he believed the policemen when they said they would keep an eye on the house. The time came when he had to go to work, but he could not until he made sure Beverly locked all the doors, and he made her promise that she would not, under any circumstances, unlock any door. She also had to promise to call him immediately if she saw or heard anything or got any strange phone calls.

He drove unwillingly to the Michoud Assembly Facility that day only to find all the engineers scrambling to find parts for Boeing's 74-foot-long, 260-inch diameter booster, Saturn V. The problem had come about because NASA had decided they wouldn't use Chrysler's much-tested, very reliable Saturn I to test the Saturn V. Yes, like the Redstone before it, they would continue to build Saturn Is, but they needed much larger Saturns to test the biggest one of all, Saturn V, and by larger they also meant Saturns II, III, and IV. All hands were on deck; the Chrysler side of Saturn Lane was scrambling just to be able to accommodate the changes in design coming down the pike for those Saturns. They had to build, finish, and perform preliminary tests on Saturns II through IV before they could even think about being ready for the greatest booster of all, which of course was Saturn V, which they already knew would take Apollo to the moon.

Saturn V lay more or less quietly on Boeing's floor, partly waiting for parts and things, but mostly because they had hit a snag. For weeks now, Boeing's men had been trying to tool Saturn V's giant ring into a perfect round, but their tools could not seem to be able to handle the work. They tried everything. The Saturn V was the most important ring of all because it would be the ring on the bottom that would hold the huge engines in place. It had to be perfectly smooth and perfectly round so as not to be knocked off course by even the slightest of challenges out in space. Even the challenges on the ground never seemed to end. Like Saturn I, if Saturn V lay in one position on the floor too long, it would lose its shape, so the workmen had to rotate the precious, much heavier giant constantly, day and night. But they had to round the ring before they could even begin to think about other problems like re-entry.

NASA, meaning Dr. von Braun, prodded Boeing to finish. He wanted it tested; he wanted every part of every missile to be tested, even the smallest parts. So much had to be done before he could consider his "brawny booster" ready to send a man to the moon. He was impatient with delays; the Soviets had sent an astronaut into orbit before he had. Dr. von Braun grumbled; that would never have happened had the government not made him test his Jupiter a second time. It had already passed one test, and if he hadn't had to send Ham, the famous rhesus monkey, into orbit for a second test, he was sure he would have beaten the USSR.

Now the Soviets were boasting about Yuri Gagarin going into orbit. Americans had done nothing but send scientific payloads into outer space, and very few payloads at that. The only nose cones America had sent into space only carried instruments designed to decide whether or not the missile that was carrying them was effective.

There was another problem. The Marshall Space Flight Center's testing facilities were not big enough to handle the larger Saturns. They all had to travel to Cape Canaveral to be tested. The Cape's Launch Complex 34 and Pegasus stayed busy.

By November 1964, Saturn V was still lying on the floor even though more than a dozen of the larger Saturns had passed their tests at Cape Canaveral. And when 1965 rolled around, Saturn V was still the holdup.

Huntsville's designers were not the only ones egging Boeing along. Several of the astronauts were also waiting for Saturn V, without success. Off and on during the years since Alan Shepard crewed America's first mission into space, Shepard would stop by the Michoud facility to see how Saturn V was coming along. Because NASA had promoted him to "Chief Astronaut" or "Chief of the Astronaut Office," Shepard was responsible for scheduling and training NASA's other astronauts. As part of that training, he wanted his astronauts to have a chance to learn on the equipment at MAF, especially on the Saturn V.

The day Shepard arrived at the Michoud Assembly Facility, Bill knew he would have to tell him the truth. Bill had long been the man responsible for visitors, and he would often take them out to dinner or to go grab a coffee, but with Alan Shepard he thought it best to explain everything while they were walking down Boeing's side of Saturn Lane.

Boeing's men had not yet been able to tool Saturn V's "ring" into a perfect round. That ring sat idly in its oil bath as half a dozen men stood by watching. Bill explained that Boeing had re-tooled an old mill for the sole purpose of rounding that ring and had moved it and all their metal work to the Louisiana facility to begin the process. They had begun with high hopes, anticipating a smooth ring

Entrance to the Michoud Operations in 1965

as part of the last step for the first Saturn V, but they had tried one method after another to complete the job. They had filled the mill with oil and submerged the ring, and used every kind of tool they could think of to machine that steel ring into a smooth round, but nothing had changed. That ring was perhaps too big or too heavy, but in any case it refused to take shape. Alan Shepard said he had not given up. He just wanted to check on the progress and find out when they thought it would finish.

"Do you have even a ball park guess as to when it will be ready?" he asked Bill.

"Well, it's close," said Bill as he toured the astronaut around Boeing's side of the floor. "They're machining the ring." He avoided saying that Boeing had been machining the ring for quite a while but had made no progress, but he did say that the ring was an essential connection for that stage of Saturn V.

"I would like to bring my astronauts by to take a look. It's important for me and for them to be familiar with it," said Shepard.

So Bill introduced him to some of Boeing's crew and encouraged him to talk to Saturn V's engineers. Boeing, he said, would be better able to answer his questions than he would. Each contractor at Michoud had its own separate space, each its own moving parts, and because Saturn V was huge, Boeing's space was huge. Boeing's Space Program took up at least sixty percent of Michoud's floor, and Bill hoped Alan Shepard's presence might encourage Boeing to finish. After his tour was over, Shepard took his notes with him and left, but not before he had extracted a promise that both Bill and Boeing would keep him abreast of their progress.

As soon as he left, Bill hopped on his scooter and rode back over to Chrysler's side. Bill was used to hopping on his scooter and riding to and from the Boeing side just to see what was going on, but going from one side of the building to another took forever. None of them could afford to waste their time. Scooters were essential. Mr. Richards, the new Chief, had bought scooters for all the supervisors at the Michoud plant, and at any given moment, dozens of scooters were running all over the place in every direction.

Boeing's side was the busiest. It had finished dozens of preliminary steps for the Saturn V, including projects for each of its five sections: engines, fins, tanks, nose cone, and the lot. Its nemesis was the thrust ring. Huntsville's designers needed to attach engines and fins to it, to surround and stabilize the fuel tanks. The ring was the most important part of Saturn V's bottom. Boeing had been trying to machine the ring in that oil bath in the mill for months, and they had done everything right, including maintaining the oil at a precise viscosity by keeping it at a constant temperature. They kept turning the ring in that mill while keeping the oil temperature steady, but the ring did not respond. Now fifty men, including all of Saturn's assembly line, stood around idle; and the longer they stood idle, the more the project fell behind schedule.

After about three months of this, one of NASA's supervisors from the Boeing side hopped on his scooter and headed down Saturn Lane and across the floor of that four million square-foot building to the Chrysler side. Bill saw him coming. The Boeing guy stopped in front of Bill's supervisor and had a chat. Then Chrysler's supervisor hollered at Bill.

"Hey Will," the fellow called him Will. "Do you know anything about that big mill?"

"Yeah, I know everything about it."

"Hop on, we're wasting time!" Bill's supervisor got on his scooter, rode over to Bill, and asked Bill to climb on. Together they rode over to the Boeing side.

Boeing's supervisor confirmed the problem. "And it's costing us! We're three months behind schedule! We can't afford it. We've got men standing around idle." Bill said he'd noticed, and Boeing's supervisor continued. "The problem is we can't get that damn ring into a smooth round. These men are sitting around waiting to mount the plate and the engines. We've tried everything, every tool in the book. The oil bath's holding its temperature and we've tried different sanders, but the thing just won't get smooth. We can't make it round."

"Will's got an idea," Bill's supervisor announced to the Boeing guy. Then Bill's supervisor told the Boeing guy to hop on his scooter, and took them all back across the floor to the other side where they could discuss it.

"You did a good job fixing up that old mill," Bill began. "And that's a helluva lot of oil in that container. But I think that's what's wrong. The tools you're using were made for working on titanium aluminum, but that ring is made of steel. That ring is the only thing on the floor that isn't aluminum. I don't know where your mill was forged, but it was forged somewhere else and it was designed for shaping aluminum. You're trying to smooth out steel in a bath that's designed for aluminum. Your oil just won't work. It's too thin."

"Thin?"

"Yes, Sir. We have to put in a better oil to support the ring. I think STP will do it."

"STP? You mean like you put in cars?"

"Like you put in cars."

"Hop on, we're wasting time," and all three men rode back to the other end of Saturn Lane to phone New Orleans and order barrels of the super lubricant. It was a huge order, but in two days the STP arrived. The men set to work sump-pumping out the old oil under Bill's watchful eye, then added in the new until they had a mix whose thickness was just right for maintaining the proper viscosity. Then they set the ring back inside its new oil bath to try again.

The operators waited some ten feet above and away from the machine, and the fifty or so men who had been standing idly by those last few weeks came closer. The operator set the heat to rise exactly as it should. There was a blast of heat, and they all watched the mercury rise. When the temperature was exactly right, they corrected the process to be sure the temperature would hold.

"Okay, Guys," shouted their supervisor. "Get your tools and start that thing turning!"

The men below got to work, and the giant responded. As Saturn V's ring began to round out, they all knew Boeing's project would move forward, and Bill and the supervisor took their scooters back to the other side.

By the time Bill got home it was dark; the work day was over. Men from the construction site, who should have been gone, still stood around leaning against their machines and smoking cigarettes as though they were waiting for someone to tell them it was time to go home. It was way past five o'clock. Those men had been there before Bill left at 7:00 a.m. There was no reason for those men to be hanging around in front of his house.

"Hey, guys, isn't it time for you to go home?" he hollered at the one closest to the driveway. Way too close to his driveway, he thought.

No answer, but the men began to shuffle away toward their equipment. Beverly and Mike had suffered from their presence all that day, still afraid to go out. Beverly told him the odor had grown stronger as the day progressed. She was beside herself with worry for Mike.

"Bacteria and viruses are bound to be floating around. I'm worried about him. He's so vulnerable, so young. Please Bill, is there anything else you can do?"

"I'm worried too, Bev, I'll fix it; I promise. They have to do something. It's one helluva health hazard!"

He'd given up on Tammany Parish's councilmen. They had listened, seemingly interested in solving the problem, and had promised they were looking into it. But the only thing he ever got was more promises. He was frustrated and angry, but couldn't make them do anything. He had long ago tried explaining to the Parish's engineers how they could fix it, but pressing the point apparently just made things worse.

"Thanks for your help. I'll check back tomorrow." Persistence was all he had left.

"Certainly," they said.

But the next morning he couldn't reach even one of them on the phone. They were either out on site or gone to lunch. He could do nothing but continue to assure his wife that both Tammany Parish and Slidell's Engineering Department were on it. But as he headed back to the Michoud Assembly Facility, he had an idea.

CHAPTER 21

Missing Boys

That night, when he drove in the driveway, once again the same two men he'd seen that very first night were hanging out in the roadway, doing nothing but resting on their shovels and watching. Bill began to wonder. Were they really employees? Really road builders? He refused to think otherwise.

The next night they were still there, but this time five or six men were walking up and down the easement, this time carrying something Bill had never expected to see. On those signs was written, in bold black letters, "Yankee go home!" That was all Bill needed to know. The South certainly had its extremists. Some Yankee boys who had come south to help register people to vote had just gone missing in Mississippi.

Beverly was waiting inside, pacing up and down with Mike in her arms. If he had ever had any illusions that they had mistakenly cut into that sewer line, he no longer did. The next morning, first thing, he dialed St. Tammy Parish's Engineering Department to give them one last chance. The secretary answered.

"Mr. Brosco, they are working on it. I can assure you; but they're not here today. May I take a message?"

"Yes, they just have to look. It's been raining and the water is rising; sewage is about to creep up my front steps. Please ask them call me as soon as they get in." As if he needed assurance, he opened the front door to find the sewer water right at the bottom step: smelly swamp water with sewage, maybe even snakes and alligators. He had to do what he could to fend it off and began digging a ditch across the front yard to divert the sludge elsewhere, but his house was built on flat land. That water would keep on rising.

The next day, he caught one of the engineers in his office. The man assured Bill that they knew about it but they hadn't been able to do anything about it because of the rain. They couldn't do anything until the rain stopped.

"Sorry, Sir. We ain't had a chance. It's just too wet."

Bill was ready to punch him but decided against it. It was now time to go one step higher. He would go in person to see Louisiana's sheriff. He was steaming mad but didn't want Beverly to worry. They didn't sleep that night or the next morning; he tried his best to soothe his tired wife, telling her this was a temporary situation. A crying baby didn't help. Whatever he had to do, he had to do it before they expected him at Michoud.

"Bill, I'm afraid," she said as he walked out the door. "Mike just threw up." He thought a minute before he gave his answer.

"Don't open the curtains. I'm going to get to the bottom of it now," he said as he left. For the second time in his life, his was afraid he might have made a promise he might not be able to keep. He jumped in his Chrysler, geared it down, then took off going over the speed limit to make it to the Louisiana sheriff's station before all the sheriff's men were out on the road. If he didn't, he would no doubt encounter a secretary of some sort who, like all the others, would just ask him to fill out another complaint. If that happened, he would have to involve a lawyer, but he didn't know any lawyers in New Orleans, much less in Slidell.

The sheriff was still in his office. When Bill asked him if there had been any progress on the arsonist who tried to burn down his house, the sheriff said they had a few suspects, but nothing concrete. Then Bill told him the rest of the story, all he had endured since the highway department dug that ditch off of Highway 190, and the sheriff nodded his head and said "I know." Apparently the sheriff had been apprised of the situation, but he took notes anyway, then excused himself to make a few calls.

When he returned, he told Bill to calm down, just as everyone else had done, but did say he would send a deputy over to his house right away. Bill thanked him and said he felt a little better, then promised that he would definitely be there to meet whoever came.

Someone had heard him out, so he called Beverly before he left the man's office, telling her what the sheriff had said. Then he turned his full attention back to Michoud and the new NASA Mississippi Test facility they had just finished building down the road. Saturn V was about to have its own testing facility right across the border where Chrysler could test not only its first stages but also second and higher stages of the missile. This would make life much easier.

The announcement could not have come at a better time. The MSFC had just received orders from President Kennedy to let the Apollo program begin. This was

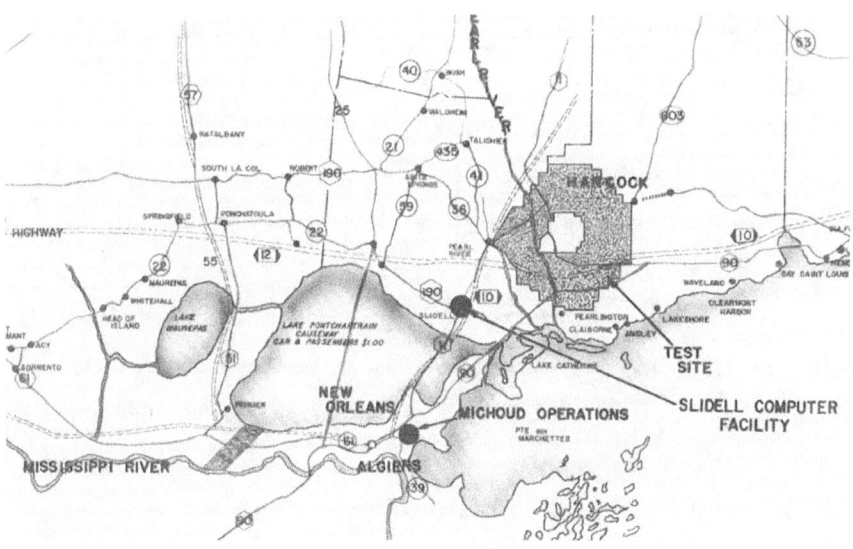

Regional map showing Mississippi Test Facility

249

progress. Kennedy wanted and was ready for the von Braun team to put a man on the moon ... now.

That afternoon, he left early and arrived home without any threatening men standing on his side of the street. The rain had stopped, which was a good sign, but the smell had not gone away. He tried playing with his son while Beverly cooked supper. It would be a good night if he could stop worrying. The sewer-swamp water had not receded enough to offset the fact that more rain was on the way.

It was clear; he no longer doubted that they would have to abandon their own house if he could not stop whatever was going on. When he'd assured Beverly that St. Tammany Parish would take care of it, he believed it; now he was lying to himself if not to her. He was the only one who could fix it, nobody else could or would. He had filled out his first report at least six weeks ago, and he'd registered complaints ever since. He had even taken Beverly and Mike with him to the Inspections and Enforcement Department and to St. Tammany Parish's council meetings, hoping to solicit a little sympathy for a mother and child. They got sympathy of course, but nothing else. He would do whatever it took.

A few days went by, and no deputy came. Bill told Beverly once again that she should not answer the door, that everything would be okay. This time he threw on his coat and stormed out the door to confront the Louisiana sheriff. There was no team to back him up, he had to do it alone. The work day had begun, and he had to park around the corner.

"Sir, this time I really need your help right away. This should be an easy fix, but it's going to take more than one man. I picked up these sketches at the office this morning. I figured they might help us figure out how to fix it." Bill handed the man the sketches. "I think you could just ask your contractors to begin by changing the slope of the rip rap. That should hold it until the job is done."

"We don't need a Yankee to tell us how to do our job. I've got your complaint."

"Sir, we have a new baby; I will do whatever I can to help you fix this problem. My wife and I would really appreciate it if you could get rid of that sewer water.

It's already lapping at my steps. It's filthy, Sir."

"I said I have your information. You Yankees want us to jump every time you say jump."

"I've filled out enough forms to break Slidell's bank and gotten nowhere," said Bill. He noted the man's name. "Andrew L. Erwin."

"You're one of those Yanks up there at the old Michoud Plant, aren't you?"

"Yes I am, and you're the sheriff for the State of Louisiana. If I'm not mistaken, the sheriff handles crimes done to property, and I've come to report a crime."

"I'm listening." But however Bill put it when he told the sheriff about his problems or property crimes, the sheriff would respond with "You don't say?"

Bill had had enough. "The Louisiana Department of Transportation has diverted swamp water into my house and I call that a crime! I'm taking it to court." Bill started to walk out.

The man slammed his fist on his desk and told Bill he called it nothing more than a construction issue. Then he asked Bill to wait right there where he was, he would make one more phone call; he knew who he had to talk to, and Bill would just have to sit there. Erwin disappeared into the radio room, where he took a lot longer than necessary for a man to make one phone call. The man finally returned. He had an answer: his deputy was finally on his way to Bill's house and would meet him there. Bill left.

Sure enough, within minutes of Bill's arrival at his house, there came a knock on the door, and when Bill opened it, he saw what he took to be a policeman in plain clothes. Unlike a policeman, this deputy wore a shirt and tie but drove an official state trooper's car.

"Come with me," was all the man said.

"Where are we going? What do you have to say? Tell me what's going on."

"I'll tell you when we get in the car, I don't think we should talk in front of your wife."

"You can tell me whatever it is, here, in front of my family." The man just frowned.

"I said come with me." The man hesitated as though he had something else to say.

Bill understood the unspoken rest of that sentence: "or else," and tried to dissuade the man one more time, then decided it best if he went along. He told Beverly he had it under control, and she was to stay where she was. He would be back in time for supper.

As the two men walked toward the cop car, Bill protested again that he would have to go to work, that they were expecting him. The man then took Bill by the shoulder and pushed him ahead of him out toward his car. He then proceeded to open the back door and shove him in.

"Let's go, Yank," said the deputy. Bill sat, dumbfounded, protesting that he wanted to know where they were going and he would take his own car. But the deputy turned the key in the ignition, and they were soon heading down Highway 190 toward what Bill hoped would be a spot where they could examine the source of that sewer water. They were getting close to the junction where Bill knew the break lay when the deputy took the wrong turn.

"Sir, there's the problem! That's it! Right down there," Bill tried to point out the levee path construction and cable barrier installations in the median, but the deputy—or whoever he was—did not respond. He just kept on driving east on Highway 190 and left Slidell in the dust.

The sun was lower in the sky now, and Bill thought about jumping out at every possible place a man might be able to jump out of a moving car. He'd tried everything, but his door would not release. He realized he'd been captive from the minute the guy shoved him into the back seat of that police car. They continued on the highway along miles of swampland until they finally crossed the Pearl River bridge. When they crossed the state line and drove on into the Covington Lowland of Mississippi, he knew exactly what was happening.

The man turned off onto a side road and into the yard of a small diner on the street side of a large field. Its neon lights read "Cafe," but from what Bill could see, there was nobody inside that diner. Outside the diner's door and in a vacant lot off to the left were several men, some sitting and others rocking back in their chairs in the shade of the awning's overhang. Just off the concrete parking area lay stacks of old tires and hubcaps; lengths of metal were piled nearby. Closer to the door was a pile of what looked like newly cut two-by-fours, but there were no signs of construction. As the three men rocked back and forth in their chairs, the awning cast shadows over them, and the neon sign blinked. The two men closest to the door squatted on their haunches watching the sheriff's car come to a stop, like jackrabbits waiting to jump.

As the deputy eased his car to a stop, Bill could see something moving inside the shop. Someone had come to the register to pay a bill and had put his cone-shaped hat on the counter.

These were no jackrabbits, and this was no mistake. Bill's tormentor stopped, opened his door and got out; then he walked around to the back and opened Bill's door.

"Stay here, Yank, if you want. I'm goin' in for a Coca Cola and I won't be long. You just stay here. If you need to go to the latrine, let one of the guys over there know and they'll show you where it is."

All of Bill's senses shot through the roof. They had crossed a state line. If something were to happen to him, he knew what the story would be. He would be the prisoner who tried to escape.

"On second thought, why don't you just come with me?" The man tried to push Bill out in front of him, but Bill would have none of it. He grabbed hold of his tormentor's coattail and hung on for dear life. When they passed the jackrabbits, Bill pulled the sheriff's coat tighter to him until he was right up against that man's back. He was no fool. The jackrabbits let them pass, then they sat back down under the shadow of the roof and pretended to be interested in the bits of lumber and metal lying around.

"Never let your opponent get behind you!" Jack's words played over and over like a stuck record in his head. The deputy continued to try to shake him off, but Bill would not be shook. The men in the shadows just sat there while Bill and the deputy went inside. When they came out together, the deputy with a Coca Cola in his hands, the jackrabbits leaned forward in their chairs. But when they saw Bill hanging on, they slunk back into their original rocker positions.

How well Bill knew that those rods and timbers were easy-to-come-by weapons: he still had a hole in his chest. He wrapped the deputy's shirttails tight in both his hands, and wherever that man moved, he moved with him. He hugged him through the cokes and through all the greetings among the men squatting by the door. Mostly, he tried to memorize all those faces. The closer they came to the squad car, the harder the deputy tried to pry Bill's hands loose. But he could not. Finally, he wrenched free from Bill's grip long enough to force him once again into the back seat of the vehicle, and as soon as he had Bill securely locked in the back, he took the wheel again and started back down the road to Slidell. The only thing Bill heard from the front seat was that what he'd hoped to find wasn't there. After that, the man mumbled that he'd made the trip for nothing.

The car headed back along the same route they'd come in on, and the deputy said he was taking him back home. Bill made mental notes of everything along the way just in case. To his relief, the deputy—who he later found out hadn't given his right name—did drive back toward Slidell, but he didn't take Bill all the way home. The deputy released him out on Highway 190, within walking distance.

When he arrived home, there were no more men hanging around in his driveway. Inside, Beverly had already put a fussy Mike down to bed. Bill kissed her and held her tight.

"I told you there was nothing to worry about." He lied of course, avoiding the details, but let her know that they had crossed the state line and that he really didn't know who the man driving was other than the sheriff had told him he was sending a deputy. Neither of them mentioned the stuff all over their yard; they had given up and were talking about selling.

The next morning, to their surprise, they woke to the fierce din of crews of men with backhoes and shovels, orders being shouted, and huge piles of dirt being moved. Bill imagined he could smell fresh air as he left the house.

Problem solved? He wondered, but all the way to the Michoud he couldn't think of anything else other than where they might go. He stayed on alert that whole day and even after he came home that night, even though the air was clear. He no longer had any doubt as to what those officials' intentions had been.

All went smoothly for a few weeks. The swamp water had indeed receded and looked as though it would not be back, so Bill cleaned up his front yard. Things seemed to be returning to some level of civility until the day he went grocery shopping with Beverly.

On that day, he and Beverly and Mike went grocery shopping together. Mike was in the buggy, and every time he got close enough to the shelves, he reached out for something so that Beverly had to keep the buggy in the middle of the aisles. Soon, Bill thought the buggy was full, but Beverly said she had been so distracted that she had to go back to pick up something she had forgotten. The buggy with Mike in it soon disappeared around the shelves toward the back of the store while Bill stood in line.

They took ages, and Bill looked at his watch. It shouldn't have taken that long just to get a can of beans. He gave her another minute then went to find her.

There she was, right where she said she would be, at the far end of the vegetable display with her back to the employees' door. A man was blocking her way, speaking to her in a low voice. As Bill got closer, he realized he'd seen that man before, and that the man who had Beverly cornered and was calling her ugly names interspersed with profanity in a low and threatening voice was one of the men who had been sitting in front of the cafe in Mississippi. Bill had had way more than enough. He grabbed the man around the neck and forced him to turn around.

"Say that to me, you xxx! I dare you!"

But Bill didn't wait for the man to answer. He balled up his fist, aimed for the man's nose and punched him hard until he fell, then he picked Mike up out of the buggy and took Beverly by the arm and quietly led them out of the store, leaving the man flat on the floor at the squash end of the vegetable counter.

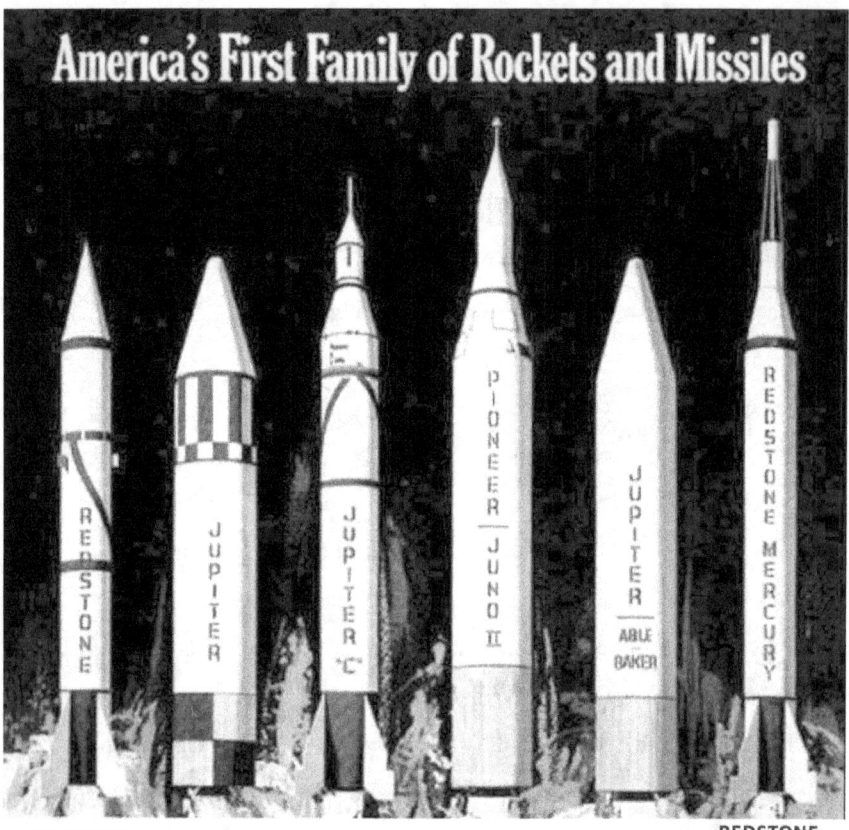

REDSTONE MISSILE
As Seen In
Warren, NH

Difference between the Redstone Missile & Redstone Rocket
The paint combinations can vary...

REDSTONE ROCKET
Launched First Astronaut,
Alan B. Shepard Into Space

CHAPTER 22

Ceremony

Years had passed since they delivered that first Saturn, but Saturns still rolled out into the world. Everything else changed: production lines, design adjustments, reinforcement of roads, constant rotation of rockets to keep them from warping, examining and re-examining even the tiniest wire or connection. The engineers worked as a team, ready for whatever, but they were plagued by the sheer numbers of demands for the new and the exacting. They made light of it to save their sanity, calling each new program today's learning opportunity, but they were and always had been a team, Dr. Wernher von Braun's team. The last thing any of them wanted was to be part of an unending, myriad, civilian, military, civic or government agency, when they were required to work together. Each of these entities had its own goals and its own resources; each wanted their project to succeed and were willing to work together, but finding common ground and trying to tap into each other's jealously guarded resources to solve a problem was not easy.

In the early days of the Saturn V-Apollo Program, only a handful of agencies and corporations had their fingers in the pie, but that handful grew to dozens and more. Even though he would not be directly involved in the Apollo, Bill knew how crucial Chrysler's part was in the next phase. He crossed himself; it had to go smoothly.

That set him thinking about his mother wanting him to be a priest. Had he

Saturn V, the Apollo Program's giant booster, waits on the Boeing side of Saturn Lane.

really ever considered it? He couldn't remember. But he thought she might be proud of him anyway, so he sat down to write one of his off and on letters to her. If she could just have seen that Saturn V booster floating down the Michoud Canal! And Dr. Von Braun had finally become a citizen of the United States! He had told her before what von Braun said every time anyone asked him if he intended to become a citizen, which was always no, because he was a citizen of the world. His becoming naturalized would be welcome news for his parents. They hadn't been too sure about their son working for a German.

Even after he became a citizen, Dr. von Braun, unlike his brother Magnus, did not leave Huntsville. When he called Bill, he always said he was planning a visit to Michoud, but he seldom came. He was too busy. NASA had made him Director of the Marshall Space Flight Center and that took all his time. But he enjoyed his new role; he was part of an even larger "team" involving many companies. and he liked to fill Bill in on what his "team" was doing.

Curry was still in Detroit, and he, too, made a point of keeping in touch with Bill, to keep him abreast of what was happening in Detroit. On the day Curry called Bill to report on one of Dr. von Braun's visits to the Warren plant, he took him by surprise. Something about von Braun, said Curry, had changed. Naturally Bill asked what, and Curry said it had to do with the Major. The Major with the gun no longer accompanied von Braun on those visits. He had disappeared. Rumors came from all quarters: maybe the Brits had him; maybe not.

Amid all the curiosity about what might have become of the Major, Bill was accused of assault. The man he'd left on the floor of the grocery store had accused him of assault, and the deputy who drove Bill across the state line served him with the warrant.

Bill went to see the judge, and the judge told him not to worry about it. He assured Bill he didn't even need to come to court, that nothing would happen.

"Your honor," asked Bill, "I've heard that you are a member of the Klan," but this didn't phase the judge. He said he'd left the Klan when he became a judge, But Bill was skeptical.

"Once a Klansman always a Klansman," he told Beverly that night, and they both knew he was right. Bill had to appear in court or the judge would declare him guilty in absentia.

The day came, and the whole family went to court. At the hearing, Bill told the judge about the deputy who took him across state lines and described his accuser as the man who had threatened him with a metal rod in front of that Mississippi cafe. The judge dismissed the case.

By 1964, when Dr. von Braun should be reveling in all of Saturn's successes, Bill knew that von Braun had turned his attention elsewhere. Those Saturns—as amazing as they were—were just one step toward putting a man on the moon. Saturn V was the "brawny booster" that would send Apollo, the capsule with the human being in it, to the moon and back. The Apollo Program was in full swing thanks to President Kennedy.

In May 1963, President Kennedy had addressed Congress and promised that the United States would send an American safely to the moon before the end of the decade. The announcement had been in response to the challenge, back in April 1961, when Americans had watched in disbelief as the Soviet Union sent Yuri Gagarin into space, and before that—in 1957—when the Soviets sent Sputnik into orbit around the earth. The United States lost to the USSR both times.

Kennedy believed that putting an American on the moon would prove not just America's superiority in space but would send a strong message to the world that Americans were doing this for the everyone, not just for the United States. He wanted his country to set the standard that no country should claim to own outer space, and that taking such a stance would keep the peace and promote freedom in the "battle going on around the world between freedom and tyranny."

He believed the United States had to take the chance, had to get to the moon before the Soviet Union. and he believed von Braun could do it. So did Dr. Wernher von Braun.

Six years after the Russians launched its last Sputnik, the space program shifted its emphasis away from military-purposed missiles to missiles as satellites designed to communicate between bases. Those satellites sent messages and control missiles both from ground to air and from land to sea. For the time, missiles no longer carried military equipment into space; they only carried scientific payloads.

Bill was acutely aware of how close they were to sending a human payload into space, as were all the contractors at the Michoud Assembly Facility, Chrysler, Boeing, and Convair—who was working on an Atlas off in the corner. The Saturn-Apollo Program went into overdrive, and NASA, under the leadership of James Webb for whom a space telescope would one day be named, expedited the tests.

Chrysler's Saturn C-1 booster had been ready for testing in record time partly because their engineers used as many existing parts as they could find, including surplus Redstone tanks and many other parts assembled by companies other than Chrysler: engines, fuel, fuel tanks, and other things. They could still test some of those parts in Huntsville, but Saturns were now far too big to be tested in Huntsville: NASA had built a new pad in Florida for that purpose.

The new facility could accommodate Saturns and all phases of the Apollo program, should they ever finish it. Saturns no longer rode part of the way over land to Huntsville, but now, after navigating the Michoud Canal and the Pearl River, they took the Intracoastal Waterway to the Gulf of Mexico and from there, they would travel around the southern tip of Florida and back up the coast to Cape Canaveral. By October 27, Saturn I was at the Cape, assembled and ready to be placed on the LC-34 pad in Block II to be tested.

Everyone watching TV and all those from Michoud, including Bill and his family, watched it lift off. It flew without a hitch, and President Kennedy was so delighted he could hardly stand it. To him, the launching of Saturns meant a new era had begun and the U.S. might finally surpass the Soviets and put a man on the moon.

It was his dream, but he would never see it happen. In November, an assassin's bullet killed him as he rode past a cheering crowd in Dallas. His dreams, though, did not die.

CHAPTER 23

Data Center

Among those who mourned for Kennedy were the women who came to every launch at Cape Canaveral—those who did the math, calculated the speeds, pressures, times, and any other computations needed before testing could even begin. They called themselves "calculators," and there were never enough of them.

NASA's designers realized it; not even thousands of calculators working together would ever be enough. To calculate the always more complicated, ever-increasing numbers of improvements and advancements like enabling an engine to gimbal safely or calculating which fuel worked best and how much fuel would be required under far different circumstances, NASA began looking at non-human calculators.

By the time Mike was five years old, NASA and Chrysler had finished building a facility specifically to house the most up-to-date, largest and most complicated computing facility in the world. The building was situated 32 kilometers, or a little less than twenty miles, southwest of the Michoud plant, which was now called the "Michoud Assembly Facility." By 1965, the state-of-the-art data center was ready to handle all engineering calculation services not only for NASA's Michoud plant but also for all of Huntsville's Marshall Space Flight Center as well as the personnel at Cape Canaveral.

The debut of this Data Center was to be the biggest event of the year. Not only were all of NASA, Chrysler, Boeing, and the rest to attend, but Dr. Wernher von Braun, after many months of absence, was about to make his first appearance at Michoud. He would attend the opening, bless Michoud's new name, and turn the Saturn program officially over to NASA. The entire State of Louisiana knew about the event, and the Mayor of New Orleans, Victor Schiro, would be the principle speaker.

Dr. von Braun's trip to New Orleans began under a cloud. The Soviets had always circulated anti-Western propaganda all over the world, but Dr. von Braun had become their primary target that year. The Soviets knew well how to create propaganda, lies skillfully linked to facts, that said things like "West Germans are raising a new Nazi army to attack us East Germans" or "Dr. von Braun is building secret ovens on missile sites to kill people."

For a long time, Wernher's friends from his days at Berlin University, which had fallen into the hands of the Soviets when the city of Berlin was divided at the end of World War II, had always managed to get mail out to him, to warn him about what the propaganda about him was saying. Wernher dismissed all of it as "lies, fabrications, and grotesque distortions."

Propaganda from the Eastern Bloc was designed to keep East Germans in East Germany. East German people were desperate to leave. The attacks on von Braun were mild compared to those the Eastern Bloc aimed at West Berlin when the Germans in the Soviet-held Eastern Bloc began starving to death. Then, when East Germans were at their weakest, the Soviets shot down an American U-2 spy plane overhead and captured its pilot.

People in East Germany were now even more desperate and were leaving by the hundreds. To stop the exodus, Stalin not only built his impenetrable Berlin Wall, but had Soviet guards manning its checkpoints. And those Soviet guards were instructed to shoot to kill. The smear campaign against everything that had anything to do with the West grew. Not only did they publish nasty articles but they distributed leaflets attacking their enemies in the West. That was when Dr.

von Braun became one of the USSR's prime targets.

People at Chrysler, including Bill, not only knew about the leaflets but some even received copies straight from the USSR. The latest rash of ugly tracts had broken out with accusations made by a Soviet "author" by the name of Julius Mader. Mader had written a book he called the "Secret of Huntsville: the true career of Rocket Baron Wernher von Braun." Those who knew about it could only shake their heads. Somehow Dr. von Braun had acquired a copy of the book.

When Dr. von Braun received his invitation from the State of Louisiana for the upcoming ceremony, he decided to make his first stop at the MAF. It had been years since he had visited. Bill would greet him as he always did; Von Braun had to see the progress on the Apollo.

The day Dr. von Braun arrived at the MAF, Bill met him at the airport as usual, but as they drove to the hotel, Bill noticed Dr. Wernher von Braun was unusually silent and he actually looked angry, or at least unhappy, which was unusual for him. Dr. von Braun hadn't even had a chance to examine anything on Saturn Lane, so it couldn't have anything to do with MAF. It had to be something else. When Bill returned the next morning to pick him up at the hotel, he said something about it.

"Good morning, Sir. Do we have bad news?" But the answer he got just puzzled him.

"Mader," he said, and got in the car.

"Mader?" asked Bill.

Dr. von Braun just shrugged.

After they arrived at the plant and when Bill accompanied the doctor down to see his brother, Wernher told Magnus everything. Bill found out who Mader was.

The next morning, Bill and Beverly dressed for the grand occasion. They had a new sitter for Mike because the last one had left because of the stench and had never come back. Shortly after the new sitter arrived, the couple kissed their son and left home for the tour of the new computer facility. They would stay for the ceremony afterward.

The tour of the Chrysler Space Division and NASA's Data Center began in Slidell, in the huge room in the center of the building that held the giant computer. Computing was what all of it was about: a mechanical computer would replace all the old punch cards. Most of the gals who had punched those cards had punched them for Wernher von Braun's team and Chrysler's aerospace engineers. The Data Center would take on many more tasks.

As Beverly and Bill toured the facilities, they spoke to each of the programmers. Bill's question was always "What are the data being fed into the computer? Is it all for the next phase of the space program?" The answers were always the same: all the data they were entering was for some aspect of the Apollo program. That computer was as important as any missile had ever been for sending man into space.

One young man, who wore a navy blue jacket with a navy-blue polka-dot tie, toured them through the satellite offices of the new facility. In the first room, he pointed out the fact that the women were still there feeding punch cards through a slot in the machine. They were the ones who would translate all the information from those punch cards into computer language—first for one of the smaller computers and then into the giant one in the main room of the building. Bill watched. All the data on the punch cards had to be translated into computer language. These women were now computing in an entirely new language.

Across the room, Bill saw something that let him know that NASA really did represent the whole world: here in the deep South, among the dozens of white female "computers" in that room, Bill saw African-Americans. One handsome young woman in a smart business suit stood out among the others. She had been intent upon what she was doing until she looked up to see Beverly walking across the room to speak to her.

"If I had still been at General Motors, I might be doing the same thing," said Beverly as she leaned over to read the young lady's badge. Beverly thanked her for what she was doing and wished her well.

They left there to go to the auditorium, where the ceremony would begin, and as they were leaving the computer building, a short gray-haired man stopped them. He had a distinctly southern accent. He introduced himself as a Federal Marshal, and he wanted to talk to Bill. The two men walked away from the crowd.

"What can I do for you, Sir?" asked Bill.

It seems that the marshal had been following the assault case against Bill. He had made it a point to be there in the courtroom when Bill was being charged with assault and had actually been following the man who charged Bill with assault.

"I just wanted to bring you up to date since your case was dismissed," he said. "We've located the car those three young civil rights workers from up North were driving before they disappeared." He drew closer to Bill and Beverly.

"Its insides were completely burned out. Nevertheless, there was plenty of evidence outside the car to lead us to a witness, and now we have several witnesses who will give enough testimony to lead us to three of the men we think killed those young men. All three are now in jail awaiting trial; all three are members of the Ku Klux Klan."

He wanted to alert Bill because he thought there were probably at least three more who should be in jail.

Bill and Beverly could hardly believe what they were hearing. The men in jail fit the description of the men Bill saw outside the cafe in Mississippi; and another one was the man who had charged Bill with assault and come to court. Bill shook the man's hand and thanked him over and over for letting him know. The two men talked a little more about the case until the marshal had to leave.

The Broscos followed the crowd out to a building closer to New Orleans that was part of the Michoud Assembly Facility and settled down in the folding chairs set up for the occasion. The mayor of New Orleans took the podium. He welcomed everyone and began a long, long speech that included a lot of Cajun humor. The audience grew restless. Even Dr. von Braun began looking at his watch. Then, just as they all thought the mayor was winding down and about to introduce Dr. von Braun, the mayor surprised them all.

He told the audience that, before the program would begin, he had a quick announcement to make. No one was as surprised as Bill was when the mayor called him up to the front of the auditorium. The mayor then went over to the podium and introduced Bill to the crowd as the "Engineer and Plant Manager at Michoud Assembly Facility," which was Michoud's new name. He held up a 9 x 10 certificate for all the audience to see and made his announcement.

"As mayor of New Orleans, I want to introduce you to the man to whom I am giving this certificate. Mr. Brosco, come on up here on stage. Here you are," he said as he handed the certificate to Bill. "It says that we in New Orleans and I as mayor have declared William R. Brosco to be an honorary Colonel in the State of Louisiana."

The audience clapped, and Bill couldn't stop smiling as he took his gift. He thanked the mayor, waved happily at his fellow workers, and returned to his seat to await the main event of the day, the reason so many had come.

The mayor took a few minutes to talk about Dr. Wernher von Braun's amazing career then asked for a big hand of applause as Dr. von Braun took the podium.

"Thank you, Mayor. I want to apologize to the audience for my accent. But of course, you know I'm from Alabama."

It was his favorite joke. He then called his brother Magnus to join him up front. Then, as a man of much dignity and as the Director of the Marshall Space Flight Center, he announced to the crowd that he would no longer be coming to the Michoud plant; he had been called by Washington to be in charge of planning the entire space program and was having to move to Washington, D.C. He assured his admirers, though, that the Apollo program would go on. He then bent graciously at the waist toward the audience as they gave him a standing ovation. It was very much deserved.

Then he asked for all present to stop a moment and remember the terrible tragedy of the first Apollo trial known only as AS-204, when fire broke out during a preflight test and cost three young astronauts their lives. Virgil (Gus) Grissom, Edward White, and Roger Chaffee were in that capsule on January 27, 1967 when the fire broke out. The capsule was filled with oxygen and it happened too fast. Everyone sat silently and remembered. Then Dr. von Braun announced that NASA had now designated that mission as Apollo 1. There would be no Apollo 2 or 3.

The first Saturn V launched Apollo 4 on November 7, 1967. Now, NASA had chosen astronaut Neil Armstrong to be the new commander. He was there among them and waved.

The second Saturn V would launch Apollo 5 in 1968. Apollo 6 would be the last uncrewed Apollo mission. Apollo 7, 8, 9, and 10 sent men into space, even into orbit around the moon, but it was not until July 16, 1969 that Apollo 11 took off for the moon.

Neill Armstrong was Commander. The other two astronauts were Lunar Module Pilot Edwin (Buzz) Aldrin and Command Module pilot Michael Collins. The whole world turned its attention toward the United States, the first to send the first human to the moon. Apollo landed on July 19, before the end of the decade, just as Kennedy had wanted ten years ago when he first addressed Congress.

Epilogue

Bill met all the astronauts and toured them through the Michoud Assembly Facility throughout the Apollo program, whenever they came. He did not work on the Apollo program. It didn't matter: just meeting the astronauts and hearing one of them say "That's one small step for a man, and a giant leap for mankind" was enough. Just being a part of that Apollo program and getting to know some of those brave and accomplished astronauts was the best thing that ever happened to him.

Everything had changed. President Richard Nixon wrote the executive order that officially, finally, removed Dr. Wernher von Braun from the Army Ballistic Missile Agency (ABMA). He would live in Washington, D.C. from now on. He would be in charge of the whole space program, but he could not take his team with him. All those who knew Dr. von Braun knew how he loved being a part of the action, the creative part, and wondered how he would survive with only a desk job. He had always said he believed in his team and wanted to be with them as the work got done, but Nixon had had other ideas.

Bill stayed on with Chrysler until Chrysler's part of the space program ended. Then he and a crew of eight began to travel for both Chrysler and other companies like TRW, Inc., who built the spacecraft Pioneer that went into orbit back in 1958. They were sent to places like Berlin and Tokyo, Mexico and Germany, even Ireland and Hartselle, Alabama, where they worked on launch vehicles, satellites, computers, and all sorts of building projects.

That was more traveling than he would like, and Mike was about ready to go to high school anyway, so it was time to move on. They looked for a place near a high school with a lot of land, perhaps enough to have a farm with a few cows or even enough for a plane and a landing strip. Bill had always wanted an airplane.

Then he heard about an opportunity back up north in a place where land was plentiful, and took a new job with Copeland Corporation in Piqua, Ohio in 1977. He liked what he saw from the very beginning because Copeland felt more like home than any other place he'd been so far. Furthermore, the guys who recruited him were as nice as they could be, as nice as they had been in the beginning, when they were trying to recruit him.

One of the other engineers at Copeland, whose name was Branson, seemed like an easy-going enough fellow, so Bill invited him and his wife over to their house. They'd been in Korea at about the same time as he had, as it turned out. They'd both flown P-51 Mustangs, and they talked about the 35,000 casualties from that war, which made Bill bring up the subject of Bob. But Branson was never in Po Hang, and he never knew anyone named Johnson. He did know the little town of Hennipen where Bob had lived; he knew about its locks.

Bill settled down into something he had always heard of but had never experienced, something he could only describe as what must be "normal." Since he had never experienced normal, it wasn't normal for him; it was better.

One night, when he was home reading the newspaper after supper and Mike was doing his homework, he turned to the advertisements page. There it was! An F-51 Mustang for sale.

"Hey, Beverly! Come here and take a look!" He had interrupted her from whatever she was doing in the kitchen.

"Give me a minute," she said, but it was a short minute. He showed her the ad.

"What would you think if I bought it?" She said she didn't know anything about flying but it was fine if he had the money, but she didn't sound too enthusiastic. He went on.

"Well, what if I don't buy the whole thing? Maybe Branson will go halfings. What do you say to that?"

"That sounds like a better idea. Remember, we bought this lot to have a little farm, you know—cows and chickens and things—we never talked about an airfield." Bill thought about that all night, then the next day he made a plan.

"Hop in the car," he told them. It was a Saturday in early Spring. "We're going for a ride in the country, and if we see any cows we're going to take a look!" The family rode out onto a long country road that ran through the lush countryside until they came upon a farm with cows ranging around inside a fence. There was a sign on the fence that said "For Sale," which could have meant either the cows or the whole farm.

Mike was ecstatic. He begged his father to stop and ran out to look over the fence at a large herd of beautiful, cream-colored Charolais cows chewing grass and nursing their calves. He'd been promised farm animals since he was a little kid. Nothing would do but all three of them get out of the car and go lean over the fence to look.

"That one! The baby one over there," shouted Mike pointing to a calf nudging its mother.

"Slow down, son. It's perfect, but you might have to let that one grow a little. Furthermore, we can't have cows without a fence. We can't let calves wander all over the city," then he promised Mike that if he helped build a fence, he would bring Mike back to this same farm to look at the calf again. Anyway, they needed to give the calf some time; it still needed its mother. That satisfied Mike until they drove away from the farm and he had other ideas.

"Dad, if we buy a calf, it will need its mother, so we have to buy a mother cow, too!"

Bill thought he was safe to agree with Mike, so he agreed. He was at least safe until they built a fence. When Beverly made some comment about not having enough room for an airstrip and cows, Bill didn't ask which one Beverly would omit. For now, there was no fence; he would worry about Beverly's choice later.

As soon as they arrived home, Bill called Branson, and the two men called the number in the ad. Before they knew it Bill and Branson were co-owners of an old World War II Mustang F-51 that needed a lot of work. The first problem they had was who would take it home and where would they put it. Bill agreed to let Branson take it to his place, at least for now.

The second problem presented two choices: come get it and fly it home or take it apart and ship it. One look at the thing and there was no longer a question. It was not ready to be flown, so they took the wings off and shipped it to Ohio, where they managed to carve out a piece of land toward the back of Bill's farm to work on it. They were two happy veterans.

The day the two men first flew it was a beautiful day, and they spent all of it tinkering and testing in every way imaginable. Every instrument on that panel had to be perfect, and every part of the cockpit. They even took the engines apart to look for the tiniest of faults, temperature, radio checks, then up in the air for slow dives, hard landings, and fast corrections until they finally declared it perfect. Bill and Mike began a fence and thought there was still plenty of room for cows.

Beverly had yet to take a spin in the Mustang, so he decided to call the farmer who owned the Charolais to see if he knew a place he could land his plane nearby. The farmer did, so he decided that this was the perfect time to take Beverly on her first flight. The three of them would take the Mustang on the long-awaited trip to that farm. Then Beverly reminded him they could not bring a cow home in a plane. The inaugural flight would have to wait until after the cow trip, so they borrowed Branson's truck and drove back out into the country.

The farmer was expecting them, and he toured Beverly and Bill all over his field and through his barn, always pointing out bits and pieces of equipment Bill would need on a farm if he wanted to raise cows. The farmer offered to sell him his milking machine among other things. Bill admitted that having such tools would make raising cattle much easier than it was back in his day. He also had to admit to himself that through all those years he had been dreaming of flying, he really had thought about cows and farming. It had always been a part of who he was.

Mike was outside walking among the Charolais, acting very mature. Beverly, too, was falling in love with the cows and had already picked out a couple of them as her favorites, as did Mike. Mike had the final say.

"Ok, Mike, which one do you want?" asked Beverly.

"We want the healthiest one, don't we?" asked Bill.

"Then that one over there, the little one with the mother. Don't you think they look more active than the others?" Mike made a good point, and in a moment of largess, Bill bought both cow and calf. From that day forward, Big Moo and Little Moo were part of the family.

The Mustang was another issue. The day came when the Mustang, which had been tested many times over and now sat immobile in the sun, was ready for its longest mission. It was time for Bill to invite Beverly on a long-awaited vacation as his passenger on the plane. They would fly together to Pensacola and enjoy Florida's Emerald Coast. That, he told her, would just be the beginning; they would have many more such vacations.

Only then, after all these many years, did Beverly admit something she had always been able to avoid. She was terrified of flying! She didn't even want to sit in a plane. Bill would never have guessed such a thing, and his heart sank. There would be no flights to Florida, at least not with Beverly.

After that, he flew the Mustang alone a few times—up over Sault Ste. Marie and Mackinac Island, occasionally with Branson or Mike, but it was never the same. As time went on, he hardly ever flew and realized he probably couldn't keep it. It was just too expensive to keep up, and he knew the hard part would be telling Branson, but he had to. After several weeks and much discussion, Branson said he, too, wanted to keep it but he, too, had to be sensible. Keeping a plane was far too expensive for one person to keep up alone.

Jointly, they advertised it for sale in several newspapers at a fair price, and it wasn't long before a seasoned pilot from Texas bought it. Bronson flew the plane to Texas. Bill drove. As all three men were there signing papers, the buyer told both Bill and Branson that he understood their feeling of loss and would make them an offer. Because he would be the new owner and because he belonged to a group of people who liked to fly, he could assure Bill and Branson that whenever they were in the neighborhood, he or another member of his group would happily take them up for a ride. He was one of a group of men and women who not only collected old planes but kept them in shape so they could fly them. They called themselves the Commemorative Air Force. Bill and Branson then left the plane in Texas and drove back to Ohio saying they would return..

Back on the farm, Beverly and Mike took very good care of Big Moo and Little Moo, so that Little Moo grew up healthy and strong. But as Little Moo grew, the family faced another devastating choice—their cows were pets! Should they keep them as pets, with one a not-so-good source of milk and the other a full-grown bull, or turn them both into steaks?

There was no choice. They could not do either. So Beverly asked around and checked the newspapers until she found a farmer who sold both milk and cheese. They approached the farmer, and as the farmer offered to buy both Big Moo and Little Moo, he made them a promise. He promised the Brosco family that he would milk Big Moo faithfully and let Little Moo grow up to be the bull he was intended to be. The three Broscos watched their pets leave in the other man's truck, and turned to go home with tears in their eyes.

That night, after the three finished supper, all Bill wanted was to make his family happy. It would soon be time for Mike to go to college, and they had neither cows nor planes to worry about. He just wondered, he asked them both, if they would like to take a car trip down to the Gulf Coast to go fishing and visit his old friends, Beavers and Dewey Destin?

Acknowledgments

B ecause I have always been intrigued with the stories people tell, story tellers are definitely among my heroes. Bill Brosco, my friend for many years, was such a story teller. The day he told me one particular story about the things he built and the programs he helped create, he ended by saying something that set me writing the rest.

"I think I am the last man alive who worked with Wernher von Braun on his first Redstone."

This is his history, and the Redstone is just one of the "Bill stories" in the book. None of it would have been possible without the help of Bill's son Michael, who was always at his father's side and who exhibited infinite patience with me as I wrestled to do justice to the histories of the Cold War and the technologies of the space race. I am in Mike's debt for tracking down his father's scrapbook and preparing a Zoom session so Bill could participate from his retirement home.

I am forever in debt to my own wonderful children, Keith, Susan, and Walter Johns who have always been my mainstays. They are my support and help, in every way.

Among those who took a personal interest in the subject and shared their knowledge with me, Bernard Arbic of the Chippewa County Historical Society was most delightful. For editing the first draft and keeping me on my toes, my thanks go to my long-suffering friend, Sylvia Alf. For giving his time to read the manuscript and help me make better sense of the more technical issues, my sincerest thanks go to my engineer friend, Bob Trefry. Christine Horner is another one I count on, and she has produced another great cover for my book, and without Shawn Wright, this book might never have come to pass. Thank you, one and all.

To all those I hope have forgiven me for failing to read their texts or for showing up late or on the wrong day, during those months I was pacing the floor trying to pull Bill's story together, I say thank you. And for you, Kind Reader, thanks for reading my book. I hope you will enjoy it and will always allow your curiosity to lead you where it will.

Photographic Credits

Page

77   Books of the U.S. Revenue Service/Coast Guard, 19th and 20th Centuries. U.S. Coast Guard Historian's Office.

90   Gilberg. Eskimo Doctor.

160   U.S. Army, Redstone Arsenal Historical Information. history.Redstone. Army.mil/ihist-1953.html

174   Rebecca Hitt. Madison County/Huntsville Convention and Visitors Bureau. www.huntsville.org/blog/list/pot/miss-baker-huntsville-monkey-who- went-to-space

197   Getarchive LLC. NASA image.

198   Chrysler Corp., Jupiter & Redstone Missile Production 1950. Periscope film.com. Detroit Historical Society, NASA image.

202   Akens. Saturn Illustrated Chronology (1968)

222   Akens.

233   NASA.

242   NASA.

249   Akens.http://apolloproject.com/sp4029/Apollo_18_48_bibliography.htm

256   Akens. Saturn Illustrated Chronology (1968) NASA.

258   Akens.

Bibliography

Air Armament Center. (2007). *History of Eglin Air Force Base.* U.S. Air Force. https://www.eglin.af.mil/Portals/56/documents/history/AFD-141104-075.pdf

Akens, D. S. editor. (1968) *Saturn illustrated chronology: Saturn's first ten years, April 1957 through April 1967*, MHR-5, George C. Marshall Space Flight Center, National Aeronautics and Space Administration. http://apolloproject.com/sp-4029/Apollo_18-48_Bibliography.htm

Ashcroft, B. (2013). *We wanted wings. A history of the Aviation Cadet Program.* Headquarters Air Education and Training Command Office of History and Research. Military Bookshop. http://media.defense.gov/2015/Sep/11…/ AFD - 150911 - 028.pdf. Accessed 15 June 2019.

Author unknown (1958) *Science: a look at man's planet: 4 minute read.* Time: the weekly newsmagazine. Vol. LXII, No. 4. https://time.com/archive/6801665/science-a-look-at-mans-planet/

Berkner, L. V. (1957). The International Geophysical Year, 1957–1958: A pattern for international cooperation in research. *Proceedings of the American Philosophical Society*, 101(2), BOX 61-76. Lloyd V. Berkner papers, Manuscript division, Library of Congress, Washington, D.C. https://lccn.loc.gov/mm82054403

Bilstein, R. E. (1980), *Stages To Saturn: A technological history of the Apollo/ Saturnl launch vehicles,* Scientific and Technical Information Branch, National Aeronautics and Space Administration, (NASA SP-4206). http://apolloproject.com/sp-4029/Apollo_18-48_Bibliography.htm

Butler, C. (1999, April 15). *NASA Oral History Project: Ballistic Missile Development Pioneers. General Bernard A. Schriever NASA Oral history.* Georgetown, Washington, D.C.

Chrysler Corporation. Space Division, (1964) *This is your Chrysler Saturn*

story, UAH Archives, Special Collections, and Digital Initiatives, accessed
January 5, 2025, Box 10, Folder 19. http://libarchstor2.uah.edu/
digitalcollections/items/show/166

Cuba: Security, Missile Crisis: Khrushchev correspondence, 1962: 23
October-19 December. (1962) Papers of John F. Kennedy. Presidential
papers. President's office files. Countries: Cuba: Security, Missile Crisis:
Khrushchev correspondence 1962: 23 October-19 December. Box 115.
Identifier JFKKPOF-115-010

Dawsey, J. Ph.D. (2019, September 10). *Wernher von Braun and the Nazi rocket
program: An interview with Michael Neufeld, PhD, of the National Air and
Space Museum.* Interview August 2019. Jenny Craig Institute for the Study
of War and Democracy. The National WWII Museum. New Orleans.
https://www.nationalww2museum.org/war/articles/wernher-von-braun-
and-nazi-rocket-program-interview-michael-neufeld-phd-national-air

Dempsey, E.J. (27 June 1951) *Project Blue Book, 1947-1967. Intelligence spot report
(unclassified 27 June 2017 SS Monrovia.* Headquarters Westover Air Force
Base, Mass. NARA T1206. https.//www.fold3.com/image/7007867, original
data from the National Archives http://www.archives.gov

Dennison, J.W. A film worth watching: Eyes of the North. Aerospace Historian,
Vol. V8, No. 1. Air Force Historical Foundation (Spring 1968).

Emme, E. M. (1965). *A history of space flight.* Holt, Rinehart, and Winston,
New York.

Gilberg, A. (1949). *Eskimo doctor* (K. Elliott, Trans.). W.W. Norton & Company.
First Danish edition, 1949.

Green, C. M., & Lomask, M. (1970). *Vanguard—a history* (SP-4202n).
National Aeronautics and Space Administration, Scientific and Technical

Information Division, Office of Technology Utilization. https://www.nasa.
gov/wp-content/uploads/2023/03/sp-4202.pdf

Greicius, T. and Hartono, N. (eds.) (2021) *JPL dares mighty things: reaching for
the stars since 1936.* CL#:21-0018. Jet Propulsion Laboratory, California
Institute of Technology. https://www.jpl.nasa.gov/timeline/

Greicius, T. and Hartono, N. (eds.) (2023) *History: JPL's beginnings and
historic timeline view a list of JPL firsts.* National Aeronautics and Space
Administration, Jet Propulsion Laboratory; California Institute of
Technology. Retrieved January 2025. www.jpl.nasa.gov./who-we-are/
history. See also documentary series: https://www.jpl.nasa.gov/who-we-are-
documentary-series-jpl-and-the-space-age

Haulman, D. L. (2003). *One hundred years of flight 1903–2002* (Air Force
Centennial Flight Commemorative ed.) Air Force History and Museum
Program at Air University Press. Maxwell AFB, Alabama.

History of the Redstone Missile System, Army Missile Command, Redstone
Arsenal, AL (unclassified) 15 October 1965. Standard Form 298 (Rec. 898).

History of Eglin Air Force Base, Air Armament Center, United States Air Force
(2007). https://www.eglin.af.mil/Portals/56/documents/history/AFD-
141104-075.pdf (accessed 9/12/2022)

Holle, M.-L. (2019). *The forced relocation of indigenous peoples in Greenland—
Repercussions in tort law and beyond.* Copenhagen Business School.
Copenhagen, Denmark. https://papers.ssrn.com/sol3/papers.cfm?abstract_
id=3492655

Invasion of Poland. Wikipedia: Today's featured article/July 19, 2005 (2005). https://
en.wikipedia.org/wiki/Invasion_of_Poland

Kennedy, R. F. (1969). *Thirteen days: a memoir of the Cuban Missile Crisis.* First
ed. Harold McMillan. Robert F. Kennedy Library; California Institute of
Technology. Pasadena, California.

Key, Maj. General M.E. Office of the Assistant to the Secretary of Defense

(Atomic Energy). (1978). *History of the custody and deployment of nuclear weapons (U): July 1945 through September 1977* (Document 64414). National Security Archive, George Washington University. https://nsarchive.gwu.edu/document/19675-national-security-archive-doc-0

Killian, K. T. (1939, October 16). *Keller and Chrysler.* Cover of Time: the weekly newsmagazine (credit: Peter A. Nyholm) see also Motors: K.T.

Log books of the U.S. Revenue Service/Coast Guard, 19th and 20th Centuries. USGC Westwind.

Medaris, Maj.-General J.B. *Countdown for decision.* (2021) Hassell Street Press. Kindle Book.

Milestones in the history of U.S. foreign relations. Office of the Historian, Bureau of Public Affairs. U.S. Department of State.

Miller, E. C. (1972) *Arctic prowlers.* Sea Classics Magazine. Westwind p. 28.

Muenstermann, L. (2024, April 5). *Meet Miss Baker—The famous squirrel monkey who made space history.* Birmingham Now. https://bhamnow.com/2022020.4/05/06/meet-miss-baker-the-famous-squirrel-monkey-who-made-space-history

Muir-Harmony, T. (2020). *Operation Moonglow: A political history of Project Apollo.* Basic Books. Hatchett Book Group, New York, NY.

NASA Information Summaries (1985) *The Early Years: Mercury to ApolloSoyuz,* PM 001 (KSC), National Aeronautics and Space Administration. http://apolloproject.com/sp-4029/Apollo_18-48_Bibliography.htm.

Neufeld, M. J. (2012). *Smash the Myth of the Fascist Rocket Baron: East German Attacks on the Wernher von Braun in the 1960s. in Imagining Outer Space: European Astroculture in the Twentieth Century,* edited by Geppert, Alexander C. T., 106–126. New York: Palgrave, Macmillan.

Neufeld, M.J. (2007). Von Braun: Dreamer of Space, Engineer of War. Portions modified by Smithsonian Institution. Article includes many attempts of East Germany to discredit Dr. Wernher von Braun.

North Atlantic Treaty Organization (2022, June 3) *A short history of NATO.* *Retrieved September 25, 2024.* https://www.nato.int/cps/en/natohq/ declassified_139339.htm

NavSource Online (1952): *Icebreaker ship photo index (AGB).* Service ship photo archive. USCGC Westwind (WAGB281).http://www.navsource.org/ archives/09/08/09080601.jpg

Odishaw, *H. International Geophysical Year* (January 1958). Science. Vol. 127, No. 3290. p115 https://www.science.org/toc/science/127/3290

Plocinski, J. R. (2008) *NATO commander to commander-in-chief: The influence of Dwight Eisenhower's as NATO's supreme commander in the "new look" defense policy* [Master's thesis, Kansas State University Manhattan, Kansas]. Defense Technical Defense Information Center approved for public release.

Redgap, C. (2008). *The Chrysler Corporation Missile Division and the Redstone missiles.* Allpar.com content. Orlando, FL. https://www.allpar.com/ threads/the-chrysler-corporation-missile-division-and-the-redstone- missiles.227813 and https://allpar.com/history/military/missiles.html (accessed 31 July 2019.

Redgap, C. (2008b). *Fly Chrysler to the moon: The Saturn rockets.* Allpar forum. https://www.allpar.com/history/military/moon.html

Redstone Arsenal.(1965). *History of the Redstone missile system.* Army Missile Command.

Serres, T. Let's visit the Westwind. U.S. Coast Guard Magazine. June 1953.

Strategic-Air-Command.com (1951). *SAC bases: Return of the first convoy. Thule Air Base.* Retrieved June 18, 2024, http://strategic-air-command.com/bases/ Thule_AFB.htm

The Piqua Daily Call (1977, August 8) William R. Brosco joins Copeland Corporation (1977, August 8). The Piqua Daily Call, p 2, Piqua, OH.

This is Your Chrysler Saturn Story (1964). A part of the NASA/COSD Manned Flight Awareness Program. Michoud Operations. Space Division, Chrysler Corporation. Government Field Printing Plant. New Orleans, LA.

Von Braun, W. (1991). *The Mars Project.* University of Illinois Press. Champaign, IL

Von Braun, W., Ordway, F. I. III (1976) *The rockets' red glare: An illustrated history of rocketry through the ages.* Anchor Press. Knopf Doubleday Publishing Group, Palatine, IL.

Waldman, C. (2020) *James E. Webb and the grand strategy of the moon landing: a political, administrative, and contextual analysis.* Inquiries: social sciences, arts, and humanities. Vol. 12, No. 9, p 1. http://www.inquiriesjournal.com/ articles/1804/james-e-webb-and-the-grand-strategy-of-the-moon-landing-a-political-administrative-and-contextual-analysis. [copyright to come]

Woods, W.D., O'Brien, F., Smeaton, W. (2002-2004) *Saturn AS-205/CSM-101 Postflight Trajectory,* Chrysler Corporation Space Division (TN-AP-68-369) (NASA CR-98345) (NTIS N92-70426). Apollo Flight Journal, http:// apolloproject.com/sp-4029/Apollo_18-48_Bibliography.htm. https://www. nasa.gov/history/afj/ap08fj/01launch_ascent.html

World on the Brink (October 1962). White House Audio Collection. Radio and television remarks on the dismantling of Soviet missile bases in Cuba. Unbeknownst to almost all the participants, JFK recorded those White House meetings including about 43 hours of secret recordings related to the Cuban Missile Crisis.

United States Coast Guard. (2020, March 27) *USCGC Westwind, 1944* (WAGB 281); ex-Severni Pulius. U.S. Department of Homeland Security.